MULTIDIMENSIONAL NONLINEAR DESCRIPTIVE ANALYSIS

Shizuhiko Nishisato

CRC Press
Taylor & Francis Group
Boca Raton London New York

CRC Press is an imprint of the
Taylor & Francis Group, an **informa** business
A CHAPMAN & HALL BOOK

MULTIDIMENSIONAL NONLINEAR DESCRIPTIVE ANALYSIS

Shizuhiko Nishisato

CRC Press
Taylor & Francis Group
Boca Raton London New York

CRC Press is an imprint of the
Taylor & Francis Group, an **informa** business

A CHAPMAN & HALL BOOK

CRC Press
Taylor & Francis Group
6000 Broken Sound Parkway NW, Suite 300
Boca Raton, FL 33487-2742

First issued in paperback 2019

ISBN-13: 978-1-58488-612-9 (hbk)
ISBN-13: 978-0-367-39064-8 (pbk)

**Visit the Taylor & Francis Web site at
http://www.taylorandfrancis.com**

**and the CRC Press Web site at
http://www.crcpress.com**

To

Lorraine, Ira, Samantha

and

In memory of George William Ford

Preface

This book is intended for those involved in data analysis in diverse areas of research. Unlike in a well-controlled and well-designed statistical experiment, many of us face data to which the notion of "data being a random sample from the normal population" does not apply. "Normal distribution" means that data must be continuous, but most data we deal with in the social sciences are categorical or non-numerical. Furthermore, we know that the relation between two normally distributed variables is by design linear. In practice, however, we encounter many nonlinear relations such as "the strength of the body is generally a concave function of age." Such a phenomenon exists and cannot be ignored for the sake of using the normal distribution assumption.

Quantification of categorical or non-numerical data is indeed a ubiquitous problem. Yet most courses for data analysis are devoted to the traditional statistics based on normal theory which affords an elegant and sophisticated inferential framework such as a confidence interval for a parameter and hypothesis testing. The current book is intended to serve the needs for those who must face the reality of typical data analysis: data are discrete or non-numerical, not necessarily sampled randomly from a population and involve not only linear but also nonlinear relations.

The book starts with some discussion of why and how this book can respond to typical needs of data analysis. The discussion is hopefully illuminating. Then the readers are exposed to conceptual preliminaries, again in non-technical terms, and to technical preliminaries needed for data analysis in the remaining chapters. Part One of the book covers these topics, which hopefully is useful to understand data analysis discussed in subsequent chapters. Part Two and Part Three contain applications of what we call "multidimensional nonlinear descriptive analysis," abbreviated as MUNDA, to diverse data types, with each chapter being devoted to one type of categorical data, a brief historical comment and basic skills peculiar to the data types. The book will end with the chapter entitled "Further Perspectives," which presents several problems that need to be solved in the near future for the current methodology. Thus, this book is an overview of MUNDA of discrete data with suggestions for future developments. The main part of the book is similar in topics to Nishisato (1994), but the topics are discussed with much more insight than in the 1994 book, as well as with new results since then.

This book is for students in the social and biological sciences and researchers in such fields as marketing research, education, health and medical sciences, psychology, sociology, biology, ecology, agriculture, economics, political science, criminology, archaeology, geology and geography. The level of the book is intermediate, but the writing style is straightforward so that one may read the book with relative ease, and at the same time it is intended to be beneficial for well-seasoned researchers engaged in data analysis.

Work related to MUNDA started in the early years of the twentieth century in ecology and biology under the name of gradient and ordination methods. The contributions to this area by ecologists and biologists from many countries are overwhelming, and they have led to an enormous number of applications to problems in their disciplines up to the present moment. Outside ecology and biology, we see familiar formulations of such methods as the method of reciprocal averages, simultaneous linear regressions and appropriate scoring by statisticians and psychologists, starting in the 1930s and 1940s, particularly in the United Kingdom and the United States. After the 1950s, the number of publications on MUNDA and related studies increased, resulting in its wide applications, particularly in Japan and France, led by eminent researchers Chikio Hayashi in Japan and Jean Paul Benzécri in France. The research spread not only to many other countries but also to diverse disciplines. It is thus timely now to pause and review the general area of MUNDA.

Finally, a personal note. I have devoted my entire research career to MUNDA, which I coined as "dual scaling." R. Darrell Bock was my mentor at the University of North Carolina who introduced me to his optimal scaling (Bock, 1960). Also instrumental to my successful graduate work at the Psychometric Laboratory in Chapel Hill were Lyle V. Jones, Director, Masanao Toda, mentor in Japan, the late Mr. and Mrs. Arthur Ringwalt, host family, the Fulbright Commission and many dear fellow students and other faculty members. Michael W. Browne introduced me to the French correspondence analysis in early 1970s and drew my attention to the problem that the joint plot of row weights and column weights in the same space was not mathematically sound. At my retirement six years ago, John C. Gower encouraged me with the words "Life exists after retirement."

During my career, I have had the pleasure of personally knowing researchers in many countries: Austria, Australia, Belgium, Brazil, Britain, Bulgaria, Canada, China, France, Germany, Greece, Hong Kong, India, Italy, Japan, The Netherlands, Pakistan, Russia, Singapore, South Korea, Spain, Sweden, Switzerland, Taiwan, USA and The West Indies. Together with many colleagues, among others, Ross Traub, Gila

Hanna, Richard Wolfe, Merlin Wahlstrom, Leslie McLean, Donald Burrill, Vincent D'Oyley, Roderick McDonald, Philip Nagy, Glen Evans, Dennis Roberts, Ruth Childs, Tony Lam, Joel Weiss, Alexander Even, Domer Ellis, Sabir Alvi, Tahany Gaddala, Joan Preston, the late Howard Russell, the late Raghu Bhargava, the late Shmuel Avital, the late Sar Khan and the late Dorothy Horn, I also taught students from such countries as Argentine, Australia, Bahrain, Brazil, Canada, Central Africa, China (Mainland, Hong Kong), Egypt, Ethiopia, Greece, India, Iran, Israel, Japan, Malawi, Malaysia, New Zealand, Nigeria, Pakistan, Peru, Philippines, Singapore, South Africa, South Korea, Sri Lanka, Taiwan, Thailand, USA, Venezuela and The West Indies. Other than my career at McGill University in Montreal and the University of Toronto, I was a visiting professor at the University of Karlsruhe, Germany (Host Professor Wolfgang Gaul), Institute of Statistical Mathematics, Japan (Host Professor Yasumasa Baba), Kwansei Gakuin University, Japan (Host Professors Takehiro Fijihara, Masao Nakanishi, Akihiro Yagi (twice), and Shoji Yamamoto) and Doshisha University, Japan (Host Professors Shigeo Tatsuki and Hirotsugu Yamauchi). I owe all of these dear friends, colleagues and students a great deal for their many years of association and friendship.

For the current book, I have three regrets. The first regret is about the partial coverage of the general topics of MUNDA. During the literature search, I was overwhelmed by the enormous contributions by ecologists and biologists since the early part of the twentieth century all the way to date and I was aware of only some of their contributions until recently. Because of this late discovery, I could not cover their contributions in a deserving way in the current book. In addition, joint correspondence analysis (Greenacre, 1988; Tateneni and Browne, 2000) is not discussed. The departure of the procedure from correspondence analysis is comparable to that of factor analysis from principal component analysis, and as such it is very important for the development of MUNDA. The second regret is about the notation. Since the current book is a summary of what I have been working on under the name dual scaling, I used my own notation of some antiquity. This was done in spite of the fact that Michael Greenacre and Jörge Blasius have been trying hard to unify the notation for MUNDA-related publications. I found it difficult to change my notation in many of my old papers to the new one. My third and greatest regret is the fact that in spite of my final statement in my 1980 book that "Inferential aspects of dual scaling deserve immediate attention by researchers," I did not include inference-related topics in this book, notably work on confidence regions, association models (loglinear analysis) versus correlation models (i.e., MUNDA), sensitivity analysis, resampling, nonsymmetric-scaling and ordered categories. Instead, some references for these topics are provided in appropriate places.

For the publication of this book, I am indebted to Rob Calver, Commissioning Editor, Statistics, Chapman and Hall/CRC for his kind advice and encouragement, Clare Brannigan for coordinating the production process, Marsha Pronin and Takisha Jackson for finalizing editorial work, Kevin Craig for the design of the cover, Katy Smith for promotional materials, and, finally but very importantly, anonymous reviewers for important and very helpful comments. My work was supported by the Natural Sciences and Engineering Research Council of Canada for many years until 2003. I hope that this book will prove to be worthy of all those who have helped me during my active career.

Shizuhiko Nishisato
Toronto, Ontario, Canada

Contents

PART I

Background

This section contains some background information, consisting of:

Chapter 1: Motivation
Chapter 2: Quantification with Different Perspectives
Chapter 3: Historical Overview
Chapter 4: Conceptual Preliminaries
Chapter 5: Technical Preliminaries

Chapter 1 contains an explanation of why multidimensional, why nonlinear and why descriptive analysis can respond to practical needs of data analysis. Chapter 2 is devoted to different ways of formulating multidimensional nonlinear descriptive analysis (MUNDA). Chapter 3 provides an extensive historical survey of relevant literature. Chapters 4 and 5 are presented as short summaries of necessary background information.

CHAPTER 1

Motivation

This book contains a description of a family of methods for quantifying categorical data, called here "multidimensional nonlinear descriptive analysis" (MUNDA). This procedure covers such methods as correspondence analysis, dual scaling, homogeneity analysis, quantification theory, optimal scaling and the method of reciprocal averages. Let us first be clear why we want to discuss multidimensional nonlinear descriptive analysis. This framework is prompted by the desire to capture every piece of information contained in data we collect. One can say that the same desire exists for any types of data analysis, and this is correct. However, the desire is not always fulfilled once a number of assumptions are introduced into analysis, such as those of underlying distributions and types of relations between variables. When one specifies a model for analysis with a good intension for an elegant form of analysis, such a model may unintentionally act as a filter which screens a substantial amount of information out of data analysis.

The use of a model may be good for the purpose of capturing an expected kind of information from the data, but in the absence of full knowledge about the data it may result in leaving out a lot of information in the data unanalyzed. To avoid such filtering of information out of data analysis, we will present MUNDA which is free from an imposition of commonly used restrictions on the way in which data are analyzed. Although the analytical framework of MUNDA may appear primitive when we compare it with highly advanced inferential statistics, the readers will still find an ample reward by the proposed analysis in terms of what it captures. In this chapter, we will examine the meanings of the title of the book as an incentive for using MUNDA as well as for understanding what MUNDA is. The opposite of MUNDA may be termed "unidimensional linear inferential analysis" (UNLIA). Let us use these contrasting words to explain some motivation of the current book.

Table 1.1 *Eight Christmas Party Plans by Ian Wiggins*

Plan 1 (X_1)	A pot-luck at someone's home in the evening
Plan 2 (X_2)	A pot-luck in the group room (daytime)
Plan 3 (X_3)	A pub/restaurant crawl after work
Plan 4 (X_4)	A reasonably priced lunch in an area restaurant.
Plan 5 (X_5)	Keep to one's self
Plan 6 (X_6)	An evening banquet at a restaurant
Plan 7 (X_7)	A pot-luck at someone's home after work
Plan 8 (X_8)	A ritzy lunch at a good restaurant

1.1 Why Multidimensional Analysis?

We can consider unidimensional analysis and multidimensional analysis as contrasting modes of data analysis. Here we will see their differences and at least a reason for opting for multidimensional analysis. This will be done using a numerical example. Ian Wiggins, now a successful consultant in Toronto, collected data for Nishisato's scaling course at the University of Toronto some thirty years ago, and the data were reported in Nishisato and Nishisato (1994). The data are borrowed with the permission of the publisher MicroStats. Fourteen subjects participated in paired comparisons of twenty-eight pairs of eight Christmas party plans, that is, 8(8-1)/2=28 pairs (Table 1.1). Subjects were researchers at a research institute in Toronto, Canada. The possible 28 pairs of the eight plans were presented to them and the subjects were asked which plan in each pair they preferred. The data table is 14x28 with elements being 1 if the first plan in the pair is preferred to the second one, or 2 if the second one is preferred to the first one (Table 1.2).

1.1.1 Traditional Unidimensional Analysis

Let us use the above data and see the results of unidimensional analysis. Bock and Jones (1968) present rigorous statistical procedures for deriving a unidimensional scale from paired comparison data and other procedures. Thurstone's paired comparison model can be described as follows: When two stimuli X_j and X_k (e.g., two tones of different frequencies) are presented to subject i, this subject's perceptions of them, called discriminal processes, are considered random

Table 1.2 *Wiggins' Christmas Party Plans Data*

j	1111111	222222	33333	4444	555	66	7
k	2345678	345678	45678	5678	678	78	8
1	1121121	222222	21121	1121	121	21	2
2	2221212	121212	21112	1112	222	12	2
3	1111121	111121	11121	1121	222	21	1
4	2121112	111112	21222	1112	222	22	2
5	2221212	221222	21212	1111	222	12	2
6	1111111	221222	21222	1111	222	22	1
7	1111121	121121	21121	1121	222	22	1
8	1111121	121221	21221	1221	221	21	1
9	1221121	221122	11121	1121	222	22	1
10	1211222	221222	11111	1222	222	11	2
11	1211111	222222	11111	1111	222	22	2
12	2222122	121111	21111	1111	111	22	1
13	1211212	222222	11111	1212	222	11	2
14	2222121	211111	11111	2121	121	21	1

variates ν_{ji} and ν_{ki}, respectively, such that

$$\nu_{ji} = \alpha_j + \epsilon_{ji}$$
$$\nu_{ki} = \alpha_k + \epsilon_{ki} \tag{1.1}$$

where α_j and α_k are called modal processes, means, or scale values of stimuli j and k, respectively, and ϵ is a normally distributed random error. It is further assumed that ϵ_j and ϵ_k are jointly normally distributed. Then, it follows that the difference discriminal process $\nu_{ji} - \nu_{ki}$, indicated as ν_{jki}, is distributed normally with mean $\alpha_j - \alpha_k$ and variance $s^2{}_{jk}$. Then, Thurstone's decision rule is:

If $\nu_{jki} \geq 0$, Subject i prefers X_j to X_k
Otherwise, Subject i prefers X_k to X_j.

Noting that ν_{jk} is normally distributed, the probability that X_j is preferred to X_k, indicated by P_{jk}, can be expressed as

$$P_{jk} = \int_0^\infty f(\nu_{jk}) d\nu_{jk} \tag{1.2}$$

where

$$f(\nu_{jk}) = \frac{1}{\sqrt{2\pi}\sigma_{jk}} e^{-\frac{1}{2}[\nu_{jk}-(\alpha_j-\alpha_k)]^2} \tag{1.3}$$

If we standardize ν_{jk} to z_{jk} by

$$z_{jk} = \frac{\nu_{jk} - (\alpha_j - \alpha_k)}{\sigma_{jk}} \tag{1.4}$$

P_{jk} can be written after a few algebraic manipulations as

$$P_{jk} = \frac{1}{\sqrt{2\pi}} \int_{-\frac{\alpha_j-\alpha_k}{\sigma_{jk}}}^{\infty} e^{-\frac{1}{2}z_{jk}^2} dz_{jk} = \frac{1}{\sqrt{2\pi}} \int_{-\infty}^{\frac{\alpha_j-\alpha_k}{\sigma_{jk}}} e^{-\frac{1}{2}z_{jk}^2} dz_{jk} \tag{1.5}$$

Let us indicate the right-hand side in terms of the normal distribution function Φ as follows:

$$P_{jk} = \Phi\left(\frac{\alpha_j - \alpha_k}{\sigma_{jk}}\right) = \Phi(Y_{jk}), \text{say} \tag{1.6}$$

where Y_{jk} is called the normal deviate corresponding to the cumulative probability P_{jk}, and

$$Y_{jk} = \frac{\alpha_j - \alpha_k}{\sigma_{jk}} \tag{1.7}$$

where

$$\sigma_{jk} = \sqrt{\sigma_j^2 + \sigma_k^2 - 2\rho_{jk}\sigma_j\sigma_k} \tag{1.8}$$

At this stage, we adopt Thurstone's case V condition: σ_{jk} is constant for all j and k. Thus, by using σ_{jk} as the unit of the scale, we can drop it from the expression, and write the above equation as

$$Y_{jk} = \alpha_j - \alpha_k. \tag{1.9}$$

In practice, we do not have probability P_{jk}, but proportion of X_j preferred to X_k, which we indicate by p_{jk}. The corresponding normal deviate is indicated by y_{jk}, which now has the expression

$$y_{jk} = \alpha_j - \alpha_k + e_{jk} \tag{1.10}$$

where e_{jk} is the discrepancy between population normal deviate Y_{jk} and sample normal deviate y_{jk}. Our task is to estimate scale values α_j and α_k in the best possible way. If we adopt a simple procedure of the unweighted least squares method, the derived scale values are such that minimize the sum of squared discrepancies $\sum_{j<k}^n e_{jk}^2$, where n is the number of stimuli. The least-squares estimate of scale value α_j thus derived is given by

$$\alpha_j^* = \frac{\sum_{k=1}^n y_{jk}}{n} \tag{1.11}$$

Table 1.3 *Paired Comparison Proportions of Plan j Preferred to k (p_{jk})*

k=	1	2	3	4	5	6	7	8
j=1	.50	.64	.43	.50	.86	.71	.43	.57
2	.36	.50	.43	.21	.79	.43	.29	.36
3	.57	.57	.50	.43	.94*	.71	.50	.71
4	.50	.79	.57	.50	.93	.79	.50	.71
5	.14	.21	.04	.07	.50	.21	.07	.29
6	.29	.57	.29	.21	.79	.50	.29	.43
7	.57	.71	.50	.50	.93	.71	.50	.50
8	.43	.64	.29	.29	.71	.57	.50	.50

Table 1.4 *Normal Deviates (y_{jk})*

k=	1	2	3	4	5	6	7	8
j=1	.00	.59	-.29	.00	1.79	.92	-.29	.29
2	-.59	.00	-.29	-1.30	1.30	-.29	-.92	-.69
3	.29	.29	.00	-.29	3.29	.92	.00	.92
4	.00	1.30	.29	.00	2.57	1.30	.00	.92
5	-1.79	-1.30	-3.29	-2.57	.00	-1.30	-.2.57	-.92
6	-.92	.29	-.92	-1.30	1.30	.00	-.92	.-29
7	.29	.92	.00	.00	2.57	.92	.00	.00
8	-.29	.69	-92	-.92	.92	.29	.00	.00

Let us look at our data in Table 1.2 and calculate paired comparison proportion p_{jk}, that is, the proportion of Plan j being preferred to Plan k (Table 1.3). Note that the proportion of 1 was replaced by $1 - \frac{1}{2N}$, N being the number of subjects, to avoid the normal deviates of plus infinity (Bock and Jones, 1968). This quantity is indicated by *. Note also that all diagonal positions were filled with 0.50, although the pairs of the same plans were not presented to subjects. The paired comparison proportions are then converted to normal deviates, y_{jk} (Table 1.4). Under Thurstone's Case V condition, the least-squares estimates of the six scale (preference) values are row averages of the normal deviates, as shown in Table 1.5 and Figure 1.1.

Table 1.5 *Scale Values in the Ascending Order*

Plan	5	2	6	9	1	7	3	4
Value	-1.72	-0.35	-0.34	-0.03	0.38	0.59	-.68	0.80

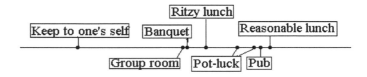

Figure 1.1 *Unidimensional Preference Scale*

In terms of this unidimensional scale, the most preferred is "reasonably priced lunch," followed by "pub/restaurant crawl," and the least preferred one being "keep to one's self." This preference scale (Figure 1.1) may appear reasonable to most eyes. However, if we look at the original data, we see a substantial amount of individual differences in response patterns, meaning that the above unidimensional scale may not reflect some or rather more than some subjects' preference orders. What happened to those individual differences?

The fact that we use the paired comparison *proportion* means that individual differences were all averaged out from the computation of the preference values of the party plans. In other words, all responses of the subjects were equally treated, and all individual differences were treated as random fluctuations. Thus, in the process of determining scale values, we have given up looking at individual differences. Those least-squares estimates of scale values have the property that they can be used to reproduce paired comparison proportions with the minimal amount of discrepancies, but they cannot generally reproduce approximations to the original paired comparison responses. The above analysis is geared towards predicting average preference values of the group, rather than preference values of individual subjects. Are we interested only in the average person's judgment? In some cases, yes, but in most cases, no.

1.1.2 Multidimensional Analysis

The same data table can be subjected to dual scaling (Nishisato, 1980a; 1994), a procedure in the family of MUNDA described in this book. In contrast to Thurstone's paired comparison scaling, we do not make such an assumption that subjects' preferences are distributed normally. Instead, subjects' paired comparison responses in hand are all that we consider for assessing scale values of the six party plans. Unlike Thurstone's method for unidimensional scaling, MUNDA is totally appropriate for the situation in which different subjects have different preferences for the party plans. Thus, generally speaking, we do not consider that there is a unique scale value for each plan that applies to everyone in the group. Instead, we consider that each plan is viewed differently by different subjects.

Taking the view that individual differences are meaningful variates, we determine a multidimensional configuration of both stimuli (e.g., six party plans) and subjects in such a way that the nearest stimulus to one subject is the most preferred stimulus for that subject, the second closest the second most preferred, and so on down to the least preferred stimulus being furthest from the subject, all of these being true to each one of the subjects. In other words, those who prefer "pub-restaurant crawl" to other plans are located closest to it, and those who prefer "evening banquet" over others are located closest to the plan "evening banquet." We can always find such a configuration of subjects and stimuli in multidimensional space. How? This is a topic for this book, and its detailed description is left for later chapters.

So, the main difference between Thurstone's unidimensional scaling and MUNDA lies in the handling of individual differences as random errors (unidimensional) or as meaningful variates (MUNDA). The analysis starts with converting the input data to a table of dominance numbers, that is, the number of times Plan j is preferred to other plans minus the number of times it is not preferred to other plans, this being calculated for each subject (Table 1.6). No assumptions are made about their distribution. From this table, we can see that Plan 5 is not much preferred to other plans and that yet it is not always the last choice for everyone. The patterns of dominance numbers over the eight plans are definitely different from subject to subject, suggesting a non-negligible amount of individual differences in the choices of party plans.

If we were to ignore individual differences totally as was done in the previous example of unidimensional analysis, one can calculate the averages of plans' dominance numbers over fourteen subjects, and the

Table 1.6 *Dominance Table*

j	1	2	3	4	5	6	7	8
1	3	-7	1	5	-1	-3	5	-3
2	-3	1	-1	5	-7	1	-5	7
3	5	3	1	-1	-7	-3	7	-5
4	1	5	-5	3	-7	-3	-1	7
5	-3	-3	1	7	-7	3	-3	5
6	7	-5	-3	5	-7	-1	3	1
7	5	1	-1	3	-7	-5	7	-3
8	5	-1	-3	1	-5	3	7	-7
9	1	-3	5	3	-7	-5	7	-1
10	-1	-5	7	-3	-7	5	1	3
11	5	-7	7	3	-5	-3	-1	1
12	-5	5	3	7	1	-7	-1	-3
13	1	-7	7	-1	-5	5	-3	3
14	-3	5	7	-1	1	-5	3	-7

average values may be adopted as scale values, in this case without any distributional assumption. In contrast, MUNDA's procedure is to differentially weight subjects in such a way that the variance of those differentially weighted averages (column means) be a maximum, one of the key objects here being to determine such weights for subjects. Once those column means of differentially weighted dominance numbers are adjusted to the proper origin and the unit (to be discussed later), they are the preference values of the party plans of Component 1. We then calculate the residual table from the input by subtracting the portion explained by the first component, and the residual table is subjected to MUNDA to determine the second component, which is orthogonal to the first component, using the same rationale for determining weights for subjects. In other words, weights for subjects this time are likely to be very different from those of Component 1. The extraction process continues until all the information associated with individual subjects is exhaustively captured.

The current example contains seven components, and the information is distributed over the components as in Table 1.7. Component 1 accounts for 34 per cent of total information (see δ), and Component 2 for 26 percent, two components together 60 per cent (see Cumδ, that is, cumulative δ). Let us look at only the first three components which

Table 1.7 *Contributions of Seven Components to Total Information*

Component	1	2	3	4	5	6	7
δ (percent)	34	26	16	13	7	3	1
Cumδ (percent)	34	60	76	89	96	99	100

account for 76 percent of the total information in data. Figures 1.2 and 1.3 are plots of Component 1 versus Component 2 and Component 1 versus Component 3, respectively. Party plans are indicated by inverted triangles and subjects by squares. In Figure 1.2, we note that the right-hand side of the graph seems to show convivial party plans as opposed to the left-hand side, which is represented by 'keep to one's self' at the extreme end. The upper part of Figure 1.2 has mostly inexpensive party plans, as opposed to expensive party plans in the lower part of the graph. In Figure 1.3, we notice that the upper side has day-time party plans as opposed to evening party plans in the lower side of the graph. In both graphs, we see that the subjects are scattered all over the space. If a graph is a 'good' representation of data, the plotted points have the following meaning: take one subject and two party plans, and the subject would choose the plan which is nearer to him or her than the plan which is further away.

It is also instructive to look at the weight of the subjects (Table 1.8). For Component 1, many of the subjects have weights close to one. If we were to ignore individual differences as in Thurstone's method of paired comparisons for unidimensional scaling, all the weights for the subjects would be 1 for Component 1 and 0 for the other components. Our results show that only two subjects have negative weights on Component 1, suggesting a modest degree of agreement in preferences among subjects who are mostly in favor of convivial parties. In contrast, weights for Components 2 and 3 are evenly distributed, indicating the subjects prefer inexpensive parties (positive weights) or expensive parties (negative weights) in an even fashion. The same is true with respect to the preferences for daytime parties and evening parties on Component 3, thus picking up a lot of individual differences. Notice that some subjects prefer an expensive evening party, some an expensive day-time party, some an inexpensive evening party and some an inexpensive day-time party.

The first component is similar to the unidimensional scale obtained

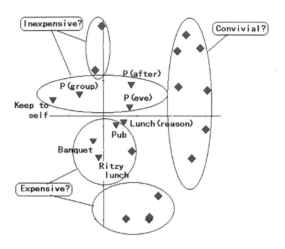

Figure 1.2 *Components 1 and 2*

earlier. However, MUNDA has shown that the first component accounts for only 34 percent of the information in data. Thus, it is difficult to justify unidimensional analysis when multidimensional analysis is possible. Generally speaking, we can safely say that we should not restrict our analysis to a single dimension, but rather be open-minded to see other possible components. We should always be prepared to deal with multidimensional information.

1.2 Why Nonlinear Analysis?

To find multidimensional structure of data, two popular methods are (a) principal component analysis when data are continuous, and (b) MUNDA

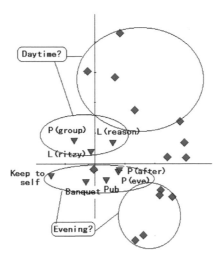

Figure 1.3 *Components 1 and 3*

when data are categorical. Let us use a numerical example of multiple-choice data. The data were first published by Nishisato (1999) and were reproduced with kind permission of the publisher Springer-Verlag. 15 subjects were asked to answer six multiple-choice questions (Table 1.9). Suppose we use the traditional Likert-type scores (i.e., integer scores for ordered categories) and regard them as continuous (no doubt there are many oppositions to this handling) for principal component analysis, that is, 1, 2, 3 as appropriate scores for the three ordered categories of each question. The data matrix is then 15x6 (The left-hand half of Table 1.10). Consider an alternative scoring scheme: give 1 if a category is chosen and 0 if not chosen. Since each question has three options and one is instructed to choose only one option per question, the response pattern of one person to one question is one of the following three, (1,0,0), (0,1,0) and (0,0,1). Thus, the data matrix is 15x12 of 1's and 0's (The right-hand half of Table 1.10). To simplify the situation, one may consider MUNDA to be essentially the same as principal component analysis of such a response pattern matrix.

Table 1.8 *Weights of the Subjects*

Subject	Component 1	Component 2	Component 3
1	1.10	0.42	-0.43
2	0.33	-1.47	1.42
3	1.21	1.16	0.10
4	0.42	-0.51	2.14
5	0.67	-1.48	0.87
6	1.50	-0.20	0.35
7	1.42	0.97	0.69
8	1.09	0.93	-0.52
9	1.53	0.37	0.10
10	0.81	-1.15	-1.17
11	1.30	-0.61	-0.55
12	-0.12	0.65	1.51
13	0.68	-1.45	-1.26
14	-0.02	1.28	-0.10

1.2.1 Traditional Linear Analysis

Principal component analysis (PCA) of continuous data is classified as linear analysis as we will shortly see its meaning. PCA is a procedure to determine a linear combination of variables in such a way that the composite score has the maximal variance. In other words, given a set of n variables, we would like to identify a single composite variable that is the most representative of all the variables. Once such a composite variable is determined, PCA moves on to seek a second composite variable that accounts for the maximal amount of information of the n variables, under the condition that it is independent of the first one. In this way, PCA will extract linear composites of variables which are independent of one another in the descending order of the variance until no more variance is left to be accounted for. PCA is a method to calculate the projections of data points on principal axes, to be discussed later. PCA typically starts with the product-moment correlation matrix, which in the current example is obtained from Likert-type (integer) scores of Table 1.10 as in Table 1.11.

The value of correlation represents the strength of linear relation between two variables. Observe a strong linear relation between blood pressures (BP) and age (Age) (r=0.66): "as one gets older the blood

Table 1.9 *Blood Pressure, Migraines, Age, etc.*

Item	Question
1	How would you rate your blood pressure? 1=Low;2=Medium;3=High
2	Do you get migraines? 1=Rarely;2=Sometimes;3=Often
3	What is your age group? 1=20-34;2=35-49;3=50-65
4	How would you rate your daily level of anxiety? 1=Low;2=Medium;3=High
5	How would you rate your weight? 1=Light;2=Medium;3=Heavy
6	What about your height? 1=Short;2=Medium;3=Tall

Table 1.10 *Likert Scores for PCA and Response Patterns for DS*

	PCA						DS					
Subject	Bpr Q1	Mig Q2	Age Q3	Anx Q4	Wgt Q5	Hgt Q6	Bpr 123	Mig 123	Age 123	Anx 123	Wgt 123	Hgt 123
1	1	3	3	3	1	1	100	001	001	001	100	100
2	1	3	1	3	2	3	100	001	100	001	010	001
3	3	3	3	3	1	3	001	001	001	001	100	001
4	3	3	3	3	1	1	001	001	001	001	100	100
5	2	1	2	2	3	2	010	100	010	010	001	010
6	2	1	2	3	3	1	010	100	010	001	001	100
7	2	2	2	1	1	3	010	010	010	100	100	001
8	1	3	1	3	1	3	100	001	100	001	100	001
9	2	2	2	1	1	2	010	010	010	100	100	010
10	1	3	2	2	1	3	100	001	010	010	100	001
11	2	1	1	3	2	2	010	100	100	001	010	010
12	2	2	3	3	2	2	010	010	001	001	010	010
13	3	3	3	3	3	1	001	001	001	001	001	100
14	1	3	1	2	1	1	100	001	100	010	100	100
15	3	3	3	3	1	2	001	001	001	001	100	010

Table 1.11 *Product-Moment Correlation Based on Likert-Type Scores*

Blood Pressure	1.00					
Migraine	-.06	1.00				
Age	.66	.23	1.00			
Anxiety	.18	.21	.22	1.00		
Weight	.17	-.58	-.02	.26	1.00	
Height	-.21	.10	-.30	-.23	-.31	1.00
	BP	Mig	Age	Anx	Wgt	Hgt

Table 1.12 *Relation of Blood Pressures to Age and Migraines*

Age	20-34	35-49	50-65	Migraine	Rarely	Sometimes	Often
High BP	0	0	4	High BP	0	0	4
Mid BP	1	4	1	Mid BP	3	3	0
Low BP	3	1	1	Low BP	0	0	5

pressures tend to become higher." In contrast, we see a very weak linear relation between BP and migraines (Mig) (r=-0.06). Let us examine these two coefficients a little further. Since each variable has three categories, the corresponding data sets can be expressed as 3x3 contingency tables, as shown in Table 1.12.

We should note that the relation between BP and Mig is not linear but strongly nonlinear (i.e., when migraines are frequent, BP is either low or high). The first two principal components (two orthogonal linear combinations associated with maximal variances) are plotted in Figure 1.4. When we draw a line going through the origin and the coordinate of an item such as BP, the line is the geometric representation of the item. Thus, three category scores 1, 2, 3 of Weight, for example, must lie on that axis (see three small dots for the three categories). If the data were collected using chronological age, for example, all subjects' ages must lie along the age axis. This restriction imposed on the so-called linear analysis represents a fundamental nature of linear analysis. It should also be mentioned that the correlation between two variables is equal to the cosine of the angle between the two axes. Thus, in the current example, we see a cluster of Age, BP and anxiety (Anx), which are linearly related to one another (i.e., "the older one gets, the higher the blood pressure" and "the older one gets, the higher the level of anxiety"). From the graph,

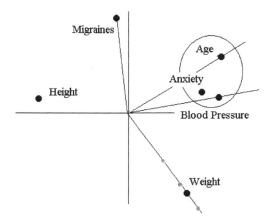

Figure 1.4 *Two Components from PCA*

it is obvious that BP and Mig are almost at the right angle, the cosine of which is close to zero.

It is important to notice here, however, that the interpretation of -0.06 being indicative of the angle between the two axes having close to 90 degrees is correct only if the two variables are from the bivariate normal distribution. In the current example, this is not the case. For a nonlinear relation, there is no geometric interpretation of correlation that $r = cos\theta$. Recall that Pearson (1904) called his correlation *normal correlation*.

The current data provide six orthogonal components, none of which, however, can capture the relation between BP and Mig because their relation is not linear.

1.2.2 Nonlinear Analysis

PCA is a special technique to extract only linear components. Thus, when data contain nonlinear relations, we wonder what PCA can do with such data. What is the linear portion of such correlation as -0.06 (the one between BP and Mig)? What does it mean when PCA decomposes -0.06 into orthogonal components?

Note that scaling of the response-pattern matrix does not seek a linear combination of items, but that of *categories* of items in such a way that the composite score has a maximal variance. Therefore, the difference between the two analyses, that is, PCA and MUNDA, may appear very slight, that is, the difference between the 15x6 matrix and the 15x12 matrix. But, they are in fact very different from each other.

We have just seen PCA being a method for linear analysis, and we now see that the scaling of the response-pattern matrix leads to non-linear analysis. Let us look at category weights associated with the first two components of MUNDA (Table 1.13).

Note that "nonlinear combinations" of response categories are involved

Table 1.13 *Projected Option Weights of Two Components*

B-PRES	Sol.1	Sol.2	MIGRAINE	Sol.1	Sol.2	AGE	Sol.1	Sol.2
Low	-0.71	0.82	Rarely	1.04	-1.08	20-34	0.37	0.56
Medium	1.17	-0.19	Sometimes	1.31	0.70	35-49	1.03	0.22
High	-0.86	-0.74	Often	-0.78	0.12	50-65	-0.61	-0.56

ANXIETY	Sol.1	Sol.2	WEIGHT	Sol.1	Sol.2	HEIGHT	Sol.1	Sol.2
Low	1.55	1.21	Light	-0.27	0.46	Short	-0.56	-0.63
Medium	0.12	0.31	Medium	0.32	0.01	Medium	0.83	-0.35
High	-0.35	-0.33	Heavy	0.50	-1.40	Tall	-0.27	0.98

in each component, except Anx and weight (Wgt). Notice also that the weights for options of BP and Mig for Component 1 are so weighted as to capture the nonlinear relation between them. Using these weights, inter-item correlation matrices are obtained for the first two components (Table 1.14).

Note that BP and Mig are now correlated at 0.99 in Component 1.

Table 1.14 *Correlation Matrices Associated with Two Components*

	Sol.1						Sol.2					
BP	1.0						1.0					
Mig	.99	1.0					.06	1.0				
Age	.60	.58	1.0				.59	-.31	1.0			
Anx	.47	.52	.67	1.0			.07	.35	.35	1.0		
Wgt	.43	.39	.08	-.33	1.0		.28	.62	-.01	.19	1.0	
Hgt	.56	.57	.13	.19	.20	1.0	.31	.29	.32	.17	.38	1.0
	BP	Mig	Age	Anx	Wgt	Hgt	BP	Mig	Age	Anx	Wgt	Hgt

As seen in Table 1.13, this was attained by assigning similar weights to high BP, low BP and frequent migraines, which are very different from the weights given to medium BP, rare migraines and occasional migraines. The same correlation for Component 2 is 0.06. Thus, notice that MUNDA involves a nonlinear transformation of each variable such that the linear correlation between the two transformed variables be a maximum.

The first two components are plotted in Figure 1.5. Unlike PCA components, three categories of a single variable are not forced to be on a single line, but usually form a triangle. It is interesting to note that PCA can never find the relation between BP and Mig and that this relation is, however, the most dominant one in MUNDA: High and low blood pressures are associated with frequent migraines. Notice further that the second dimension identifies different association between low and high blood pressures, the former with young, skinny and tall subjects and the latter with old, heavy and short subjects. Notice a remarkable

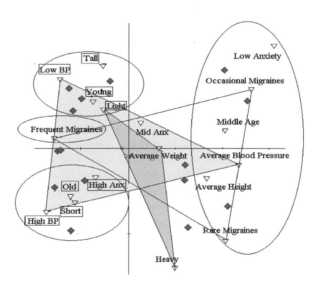

Figure 1.5 *First Two DS Solutions*

difference in graphs between PCA and MUNDA. The freedom from linear restriction in MUNDA has significant implications for multidimensional analysis, for it opens up the possibility for tapping into not only linear but also nonlinear relations between variables. Looking at the graph, we can summarize the major results as in Table 1.15.

Notice that the associations of categories indicate a number of non-

Table 1.15 *Characteristics of Two Components*

Component 1		Component 2	
One End	The Other End	One End	The Other End
Low BP	Medium BP	High BP	Low BP
High BP	Rare migraine	Rare migraine	Occasional migraine
Frequent migraine	Middle age	Old	Young
Old age group	Low anxiety	Heavy	Tall
High anxiety	Medium height	Short	
Short			

linear relations among the variables, that is, clusters of mixed orders of categories.

1.3 Why Descriptive Analysis?

In frequent applications of statistics to data analysis, we typically assume that data are a random sample from the normally distributed population. This setup is crucial for many inferential procedures. In our example of Thurstonian unidimensional scaling of paired comparison data, we assumed that the discriminal process follows the normal distribution. In that particular application, we assumed that subjects' perceived preference values are distributed normally, that is, individual differences follow the normal distribution. What happens if we are interested in individual differences? How often in the social sciences can we encounter the situation in which data can be regarded as a random sample from the normal population?

In the example of Thurstonian scaling, the scale values are obtained from paired comparison proportions, which are summary statistics in which information about individual differences is no longer contained. As we have so far seen in MUNDA, we consider individual differences not as random fluctuations but substantive variates. If we were to consider an inferential framework for the investigation of individual differences, we must design an experiment in which one person responds to the same set of stimuli repeatedly so that the fluctuation of responses from one subject may be formulated into analysis. This kind of setup is rarely used in obtaining data we typically analyze. In fact, in the majority of situations, one subject provides only one response, reducing the possibility of incorporating an inferential framework around each person, but leaving us to analyze data in a descriptive non-inferential way. It is indeed rare that we can legitimately introduce the notion of probability to describe a response of a single subject.

As we saw earlier, we should remember that the assumption of a normal distribution means that data are continuous and that the relation between variables is linear. When the normal distribution is not true, there exists information loss in product-moment correlation (e.g., Fhanér, 1967). Two conditions (data being continuous and the inter-variable relation being linear) severely restrict the possibility of an inferential framework for many data sets we deal with in the social sciences. Thus, the necessity that we must be satisfied with descriptive analysis may be a blessing in disguise.

We often talk about data analysis, of which the main aim is to analyze data in an exhaustive way, that is, without filtering out any information in data by imposing such a condition as the normal distribution. In this regard, unless the normal distribution is assumed, it is difficult to justify PCA of a Pearsonian *normal correlation* (Pearson, 1904) matrix as a means of data analysis, for under non-normal distributions Pearsonian correlation ignores nonlinear information. For example, that the correlation is 0.06 tells us either the two variables are not much linearly related, or the angle between the two corresponding vectors is close to 90 degrees. In the former case, it does not even convey what kind of nonlinear relation is involved (e.g., quadratic, cubic, quartic, random). Recall Pearsonian correlation of Likert-scored variables between blood pressure and migraine. All PCA showed was that the two variables lacked linear relation, and did not offer any pursuit of what kind of their nonlinear relation was. Pearson (1904) called his correlation *normal correlation*, telling us that it is meaningful when the bivariate normal distribution can be assumed.

Considering all of these, we cannot always afford the framework of inferential analysis, hence must resort to descriptive analysis where the sole aim is to explain data we have in the best possible way. Resampling methods such as jack-knifing and bootstrapping are feasible, but they are short of generalizing the sample results to the population unless data are sampled randomly from the population.

Nonlinear relations, multidimensional associations and individual differences are all worthwhile information in data. To capture them, we find multidimensional nonlinear descriptive analysis (MUNDA) a good alternative tool for data analysis. In the current book, MUNDA is presented as a procedure for analyzing a variety of categorical data. No doubt there are other kinds of categorical data, but we hope that the description of MUNDA in this book will provide a basic framework for analysis of general categorical data, and that the ingenuity of researchers can be incorporated into expanding the applicability of MUNDA to other types of categorical data which are not discussed in this book. Because of MUNDA's ability for exhaustive analysis of information in data, one can see a great possibility of using it for analysis of categorized continuous variables as well. This extension, however, is still at an infantile stage of development, and much more work needs to be done. This matter will be briefly discussed in the last chapter.

CHAPTER 2

Quantification with Different Perspectives

The object of this book is to introduce and discuss a family of methods for quantifying categorical data, known by such names as dual scaling, correspondence analysis, homogeneity analysis, optimal scaling and reciprocal averaging. All of these are placed under the umbrella of MUNDA (multidimensional nonlinear descriptive analysis) in this book.

This chapter is an introduction to a number of ways in which the task of quantifying categorical data can be conceptualized and formulated, thus serving as an introduction to MUNDA.

2.1 Is Likert-Type Scoring Appropriate?

Let us use a numerical example (Nishisato, 1980a) with the permission from the copyright holder Shizuhiko Nishisato. Subjects were asked the following two multiple questions and the responses are summarized in Table 2.1.

Q1: What do you think of taking sleeping pills?
(1) strongly disagree; (2) disagree; (3) indifferent; (4) agree,
(5) strongly agree
Q2: Do you sleep well every night?
(1) never; (2) rarely; (3) sometimes; (4) often; (5) always.

In the social sciences, we often use Likert-type integer scores for ordered sets of categories (Likert, 1932), which are, for example, 0 for never, 1 for sometimes, 2 for often and 3 for always. Suppose we assign -2, -1, 0, 1, 2 to the five ordered categories of each set in the above example. Our question here is if these Likert scores are appropriate and how to examine the appropriateness. There is a simple way to examine it. First, we calculate the mean of each category, using these scores. For example, the mean of category "Never" is

Table 2.1 *Sleeping and Sleeping Pills*

	N*	R	S	O	A	Sum	Score
SA**	15	8	3	2	0	28	-2
A	5	17	4	0	2	28	-1
N	6	13	4	3	2	28	0
F	0	7	7	5	9	28	1
SF	1	2	6	3	16	28	2
Sum	27	47	24	13	29	140	
Score	-2	-1	0	1	2		

*N=never, R=rarely, S=sometimes; O=often, A=always
**SA=strongly against; A=against; N=neutral; F=for;
SF=strongly for

$$[15 \times (-2) + 5 \times (-1) + 6 \times 0 + 0 \times 1 + 1 \times 2)]/27 = -1.2.$$

Likewise, we can calculate the means of row categories and those of column categories, which are summarized in Table 2.2. We now plot

Table 2.2 *Likert Scores and Weighted Means*

Score	Mean	Score	Mean
-2	-1.2	-2	-1.3
-1	-0.5	-1	-0.8
0	0.4	0	-0.6
1	0.5	1	0.6
2	1.3	2	1.1

those averages against the original scores (-2, -1, 0, 1, 2) as seen in Figure 2.1. The two lines are relatively close to a straight line, which indicates that the original scores are "fairly good."

Suppose we use, instead of those subjective category weights, those derived by MUNDA, and calculate the weighted category means. The weights derived here are called optimal weights or scores (Table 2.3). Optimal weights and weighted category means are plotted (Figure 2.2). Notice that the two lines now merge into a single straight line. This is

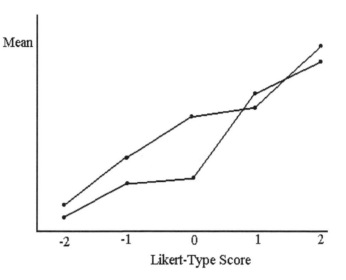

Figure 2.1 *Likert-type scores and the means*

Table 2.3 *Optimal Scores and Weighted Means*

Score	Mean	Score	Mean
-1.30	-0.84	-1.20	-0.78
-.059	-0.38	-0.64	-0.42
0.43	0.28	-0.49	-0.32
0.58	0.38	0.87	0.56
1.55	1.00	1.47	0.95

mathematically optimal, although we have not yet discussed what we mean by "mathematically optimal." We will also see shortly that the slope of the straight line is equal to the maximal "non-trivial" singular value for this data set.

But, how do we arrive at optimal weights? It is simple: Once we obtain the average category scores as in Table 1.2 and Figure 2.1,

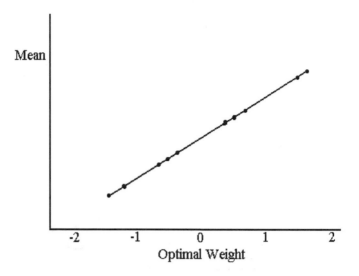

Figure 2.2 *Optimal scores and the means*

replace the original scores (e.g., -2, -1, etc.) with the corresponding mean scores, and then calculate the new mean category scores in the same way as before, using these means as weights. Plot the new category scores against the first mean scores; replace the old mean scores with the new mean scores, and calculate new mean category scores, and plot them. This is known to be a process which always converges to a straight line. In other words, the successive reciprocal averaging scheme produces a mathematically convergent sequence (See its mathematical treatment in Nishisato, 1980a, pp.60-62, 65-68).

According to Paul Horst (Personal communication, September 1970), "the late Dr. Marion W. Richardson proposed that

(1) the scale value of an item be defined as the average of the ratings of persons for whom the item was checked, and

(2) the score of a person be defined as the average of the scale values of the items checked for him."

Table 2.4 *Evaluation of Three Teachers*

Teacher	Good	Average	Poor	Sum
While	1	3	6	10
Green	3	5	2	10
Brown	6	3	0	9
Sum	10	11	8	29

Horst further mentioned that he replaced the average in (2) with a "linear function of the average," and that an iterative computational procedure, called "the method of reciprocal averages " was developed, but was not published. Horst mentioned that it was before the publication of the paper by Richardson and Kuder (1933). Historically, however, it looks as though this idea of reciprocal averaging was already known to some ecologists. It may therefore not be Richardson and Kuder (1933) or Horst (1935), but ecologists and biologists, who should be regarded as inventors of the method. The same procedure was also suggested by Fisher (1940), and fully illustrated by Mosier (1946). Let us now look at this probably first procedure ever developed for hand calculation for MUNDA, which Horst (1935) called the method of reciprocal averages (MRA). Baker (1960) then provided a computer program for Univac computers.

2.2 Method of Reciprocal Averages (MRA)

We borrow a description from Nishisato and Nishisato (1994), with the permission by the publisher MicroStats, to illustrate the procedure. Three teachers (say White, Green, and Brown) were rated with respect to their teaching performance by students (Table 2.4). The MRA is carried out in the following way:

(1) Assign arbitrary weights to columns (or rows, if preferred). Although such values are arbitrary, one must avoid identical weights for all columns (or rows) including zero. It is always a good strategy to use "reasonable" values. As an example, consider:

$$x_1(\text{good}) = 1, \ x_2 (\text{average}) = 0, \ x_3 (\text{poor}) = -1.$$

(2) Calculate the weighted averages of the rows.

$$y_1(White) = \frac{1 \times x_1 + 3 \times x_2 + 6 \times x_3}{10}$$

$$= \frac{1 \times 1 + 3 \times 0 + 6 \times (-1)}{10} = -0.5.$$

Similarly,

$$y_2(Green) = \frac{3 \times 1 + 5 \times 0 + 2 \times (-1)}{10} = 0.1000$$

$$y_3(Brown) = \frac{6 \times 1 + 3 \times 0 + 0 \times (-1)}{9} = 0.6667$$

(3) Calculate the mean responses weighted by y_1, y_2, y_3

$$M = \frac{10y_1 + 10y_2 + 9y_3}{29}$$

$$= \frac{10 \times (-0.5) + 10 \times 0.1 + 9 \times 0.6667}{29} = 0.0690$$

(4) Subtract M from each of y_1, y_2, y_3, and adjusted values are indicated again as y_1, y_2, y_3, respectively:

$$y_1 = -0.5000 - 0.0690 = -0.5690$$
$$y_2 = 0.1000 - 0.0690 = 0.0310$$
$$y_3 = 0.6667 - 0.0690 = 0.5977$$

(5) Divide y_1, y_2, y_3 by the largest absolute value of y_1, y_2, y_3, which is indicated as g_y, that is, $g_y = 0.5977$. Adjusted values are again indicated as y_1, y_2, y_3:

$$y_1 = \frac{-0.5690}{0.5977} = 0.9519$$

$$y_2 = \frac{0.0310}{0.5977} = 0.0519$$

$$y_3 = \frac{0.5977}{0.5977} = 1.0000$$

(6) Using these new values as weights, calculate the averages of the columns:

$$x_1 = \frac{1 \times (-0.9519) + 3 \times 0.0519 + 6 \times 1.0}{10} = 0.5204$$

$$x_2 = \frac{3 \times (-0.9519) + 5 \times 0.0519 + 3 \times 1.0000}{11} = 0.0367$$

$$x_3 = \frac{6 \times (-0.9519) + 2 \times 0.0519 + 0 \times 1.0000}{8} = -0.7010$$

(7) Calculate the mean responses weighted by x_1, x_2, x_3:

$$N = \frac{10 \times 0.5204 + 11 \times 0.0367 + 8 \times (-0.7010)}{29} = 0$$

(8) Subtract N from each of x_1, x_2, x_3. Since $N = 0, x_1, x_2$, and x_3

remain the same.

(9) Divide each element of x_1, x_2, x_3 by the largest absolute value

of the three numbers, which is indicated as g_x. Since -0.7010 has the largest absolute value, $g_x = 0.7010$. Adjusted values are indicated again as x_1, x_2, x_3:

$$x_1 = \frac{0.5204}{0.7010} = 0.7424$$

$$x_2 = \frac{0.0367}{0.7010} = 0.0524$$

$$x_3 = \frac{-0.7010}{0.7010} = -1.0000$$

Reciprocate the above averaging processes (step 2 through step 9) until all the six values are stabilized (Table 2.5).

Iteration 5 provides the identical set of numbers as iteration 4. Therefore the process has converged to the optimal solution in four iterations. Notice that the largest absolute values at each iteration, g_y and g_x, also converge to two constants, 0.5083 and 0.7248. Nishisato (1988c) showed that

$$\rho^2 = g_y g_x, \quad \text{and} \quad \rho = \sqrt{g_y g_x} \qquad (2.1)$$

where ρ^2 is the correlation ratio or the eigenvalue, while ρ is the singular value or the maximized correlation between rows and columns of the table. In the current example,

$$\rho^2 = 0.5083 \times 0.7248 = 0.3648$$
$$\rho = \sqrt{0.5083 \times 0.7248} = 0.6070$$

If we start with the cross-product symmetric table, instead of the raw data as in the present example, the process will converge to one constant of g, which is the *eigenvalue* and its positive square root is the *singular value* (Nishisato, 1980a). If you are wondering why the final value of g

Table 2.5 *Iterative Results*

	Iter2 y	Iter2 x	Iter3 y	Iter3 x
1	-0.9954	0.7321	-0.9993	0.7321
2	0.0954	0.0617	0.0993	0.0625
3	1.0000	-1.0000	1.0000	-1.0000
$g_{(y,x)}$	0.5124	0.7227	0.5086	0.7246
	Iter4 y	Iter4 x	Iter5 y	Iter5 x
1	-0.9996	0.7311	-0.9996	0.7311
2	0.0996	0.0625	0.0996	0.0625
3	1.0000	-1.0000	1.0000	-1.0000
$g_{(y,x)}$	0.5083	0.7248	0.5083	0.7248

is the eigenvalue, see Nishisato (1994, p.89)

(10) The unit for y_1, y_2, y_3 now needs to be equated to the unit for x_1, x_2, x_3. Following Nishisato (1980a), the unit is chosen in such a way that the sum of squares of weighted responses is equal to the number of responses. In this case, the constant multiplier for adjusting the unit of y_1, y_2, y_3 is given by c_r, where

$$c_r = \sqrt{\frac{29}{10y_1^2 + 10y_2^2 + 9y_3^2}} = 1.2325 \qquad (2.2)$$

The constant multiplier for adjusting the unit of x_1, x_2, and x_3 is given by c_c, where

$$c_c = \sqrt{\frac{29}{10x_1^2 + 11x_2^2 + 8x_3^2}} = 1.4718$$

The final weights are obtained by multiplying y_1, y_2, y_3 by c_r, and x_1, x_2, x_3 by c_c. These weights are called *normed weights* or *standard coordinates*. The normed weights, multiplied by the singular value, that is, ρy_i and ρx_j, are called *projected weights* or *principal coordinates*. The distinction between these two types of weights will be discussed later. The final results are as shown in Table 2.6. These weights are scaled

Table 2.6 *Two Types of Optimal Weights*

	normed y	normed x	projected y	projected x
1	-1.2320	1.0760	-0.7478	0.6531
2	0.1228	0.0920	0.0745	0.0559
3	1.2325	-1.4718	0.7481	-0.8933

Table 2.7 *Data Expressed in Terms of Unknown Weights*

Teacher	Weighted Response	Total
1	x_1,x_2,x_2,x_2,x_3 x_3,x_3,x_3,x_3,x_3	$x_1 + 3x_2 + 6x_3 = z_1$
2	x_1,x_1,x_1,x_2,x_2 x_2,x_2,x_2,x_3,x_3	$3x_1 + 5x_2 + 2x_3 = z_2$
3	x_1,x_1,x_1,x_1,x_1 x_1,x_2,x_2,x_2	$6x_1 + 3x_2 = z_3$
		$z_1 + z_2 + z_3 = z_t$

in such a way that (1) The sum of responses weighted by y is zero, and the sum of responses weighted by x is zero, (2) The sum of squares of responses weighted by y is the total number of responses, and the same for x. Once the first component is obtained, we calculate the residual frequencies, and apply the MRA to them to extract the second component. This process will be discussed later.

2.3 One-Way Analysis of Variance Approach

We will use Table 2.4 to illustrate this approach. Let x_1 be an appropriate weight for 'good,' x_2 for 'average' and x_3 for 'poor.' In terms of these unknown weights, the original data in Table 2.4 can be rewritten as in Table 2.7. The object of our scaling is to determine those weights which describe one teacher to be as close as possible among themselves and as different as possible from those weights which describe another teacher. This is what we call Guttman's principle of internal consistency. Let us define the following terms:

$$a = f_t = 10+10+9=29$$

$$b = \text{'correction term'} = \frac{\sum(\sum f_{ij}x_j)^2}{f_t} = \frac{z_t^{\,2}}{f_t}$$

$$c = \sum f_{.j}x_j{}^2 = 10x_1^2 + 11x_2^2 + 8x_3^2$$

$$d = \sum \frac{(\sum f_{ij}x_j)^2}{f_j} = \frac{z_1^2}{10} + \frac{z_2^2}{10} + \frac{z_3^2}{9}$$

In the context of the one-way analysis of variance, the total sum of squares (SS_t) can be decomposed into the sum of the between-teacher sum of squares (SS_b) and the within-teacher sum of squares (SS_w), that is,

$$SS_t = SS_b + SS_w \tag{2.3}$$

where $SS_t = c - b$, $SS_b = d - b$ and $SS_w = c - d$. Divide both sides of the above equation by SS_t, which yields

$$1 = \frac{SS_b}{SS_t} + \frac{SS_w}{SS_t} \tag{2.4}$$

To apply Guttman's principle of internal consistency to this case, our task is then to determine the three category weights (x_1, x_2, x_3) so as to maximize the between-teacher sum of squares, relative to the total sum of squares. This relative quantity is called the correlation ratio and is indicated by η^2, namely,

$$\eta^2 = \frac{SS_b}{SS_t} \tag{2.5}$$

It follows then

$$\frac{SS_w}{SS_t} = 1 - \eta^2 \text{ and } 0 \leq \eta^2 \leq 1 \tag{2.6}$$

Thus, the task is achieved by determining the category weights so as to maximize η^2 or minimize the ratio of SS_w over SS_t. Since the correlation ratio is invariant over the origin of measurement, we set the sum of weighted scores to be zero,

$$10x_1 + 11x_2 + 8x_3 = 0$$

which makes the correction term (b) zero as well. Then,

$$SS_b = \frac{(x_1 + 3x_2 + 6x_3)^2}{10} + \frac{(3x_1 + 5x_2 + 2x_3)^2}{10}$$

$$+\frac{(6x_1 + 3x_2)^2}{9}$$

$$SS_t = 10x_1^2 + 11x_2^2 + 8x_3^2.$$

Maximization of the three category weights is not a straightforward matter, for it involves such a *trivial solution* that irrespective of data the correlation ratio becomes one, the absolute maximum. This trivial solution represents the information expected when the rows and the columns of the frequency data are statistically independent, that is, when the joint frequency of row i and column j is equal to $\frac{f_{i.}f_{.j}}{f_t}$. Since we are not interested in the trivial solution, we first eliminate it from the data, and analyze the residual frequencies, say f_{ij}^*, where

$$f_{ij}^* = f_{ij} - \frac{f_{i.}f_{.j}}{f_t} \tag{2.7}$$

This matter will be revisited later when we discuss how to maximize the correlation ratio in terms of the category weights.

In the current example, the category weights that maximize the correlation ratio, namely the optimal normed weights for 'good, average and poor' are $1.0761, 0.920$ and -1.4717, respectively and the correlation ratio is 0.3683.

Once the category weights are obtained, the corresponding scores for the teachers y_i can be obtained by

$$y_i = \frac{1}{\eta_1}\frac{\sum_{j=1}^{J} f_{ij}x_j}{f_{i.}} \tag{2.8}$$

Using this formula, the optimal normed scores for teachers 1, 2 and 3 are $-1.2322, 0.1227$ and 1.2326, respectively. Projected scores and projected category weights are obtained by multiplying the respective normed scores and weights by the singular value, which is $\sqrt{0.3683}=0.6069$.

Before moving to the next approach, let us reformulate the above approach in matrix notation. The data matrix is indicated by $\mathbf{F}$,

$$\mathbf{F} = \begin{bmatrix} 1 & 3 & 6 \\ 3 & 5 & 2 \\ 6 & 3 & 0 \end{bmatrix}$$

$\mathbf{f_c}$ is the column vector of the column totals of $\mathbf{F}$,

$$\mathbf{f_c} = \begin{bmatrix} 10 \\ 11 \\ 8 \end{bmatrix}$$

D_c is the diagonal matrix of the column totals of F,

$$D = \begin{bmatrix} 10 & 0 & 0 \\ 0 & 11 & 0 \\ 0 & 0 & 8 \end{bmatrix}$$

D_r is the diagonal matrix of the row totals of F,

$$D_r = \begin{bmatrix} 10 & 0 & 0 \\ 0 & 10 & 0 \\ 0 & 0 & 9 \end{bmatrix}$$

y is the column vector of scores for the teachers,

$$y = \begin{bmatrix} y_1 \\ y_2 \\ y_3 \end{bmatrix}$$

x is the column vector of weights for the categories,

$$x = \begin{bmatrix} x_1 \\ x_2 \\ x_3 \end{bmatrix}$$

Three terms of the one-way analysis of variance can be expressed as

$$SS_t = x'[D_c - \frac{f_c f_c'}{f_t}]x \tag{2.9}$$

$$SS_b = x'[F'D_r^{-1}F - \frac{f_c f_c'}{f_t}]x \tag{2.10}$$

$$SS_w = x'[D_c - F'D_r^{-1}F]x \tag{2.11}$$

Since the origin and the unit of quantified variables are arbitrary, we choose the origin and the unit as follows:

$$f_c'x = 0, \qquad SS_t = f_t \tag{2.12}$$

Our task is to determine the vector of category weights so as to maximize SS_b, subject to the conditions that $SS_t = f_t$ and the sum of weighted responses to be zero. This task can be handled by Lagrange's method: Define the Lagrangian function, which is in the present case,

$$Q(x, \lambda_1, \lambda_2) = SS_b - \lambda_1(SS_t - f_t) - \lambda_2(x'f_c f_c'x - 0) \tag{2.13}$$

Partially differentiating the Lagrangian function with respect to x and two Lagrangian multipliers, and setting them to 0, we obtain

$$\frac{\partial Q(\mathbf{x}, \lambda_1, \lambda_2)}{\partial \mathbf{x}}$$

$$= 2[\mathbf{F}'\mathbf{D}_r^{-1}\mathbf{F} - \frac{\mathbf{f}_c\mathbf{f}_c'}{f_t}]\mathbf{x} - 2\lambda_1[\mathbf{D} - \frac{\mathbf{f}_c\mathbf{f}_c'}{f_t}]\mathbf{x} - 2\lambda_2\mathbf{f}_c\mathbf{f}_c'\mathbf{x} = \mathbf{0}$$

$$\frac{\partial Q(\mathbf{x}, \lambda_1, \lambda_2)}{\partial \lambda_1} = \mathbf{x}'[\mathbf{D} - \frac{\mathbf{f}_c\mathbf{f}_c'}{f_t}]\mathbf{x} - f_t = 0$$

$$\frac{\partial Q(\mathbf{x}, \lambda_1, \lambda_2)}{\partial \lambda_2} = \mathbf{x}'\mathbf{f}_c\mathbf{f}_c'\mathbf{x} = 0$$

The last two are nothing but the constraints. We can now simplify the first expression to arrive at the following:

$$[\mathbf{F}'\mathbf{D}_r^{-1}\mathbf{F} - \lambda\mathbf{D}_c]\mathbf{x} = \mathbf{0} \longrightarrow \mathbf{F}'\mathbf{D}_r^{-1}\mathbf{F}\mathbf{x} = \lambda\mathbf{D}_c\mathbf{x} \qquad (2.14)$$

This is called a generalized eigenequation. Notice that the standard form has the identity matrix, in lieu of $\mathbf{D}_c$. The value λ is an eigenvalue and the vector $\mathbf{x}$ is the corresponding eigenvector. We know that the eigenvalue is nothing but the correlation ratio we want to maximize, which can be seen by first premultiplying both sides by $\mathbf{x}'$ and then rearranging the resultant expression as follows:

$$\lambda = \frac{\mathbf{x}'\mathbf{F}'\mathbf{D}_r^{-1}\mathbf{F}\mathbf{x}}{\mathbf{x}'\mathbf{D}_c\mathbf{x}} = \frac{SS_b}{SS_t} = \eta^2 \qquad (2.15)$$

where η^2 is the correlation ratio. In this book, therefore, the correlation ratio and the eigenvalue will be used in an interchangeable way. Notice that the quantified data are centered, thus without the mean in the formula.

At this stage, it is convenient to introduce a new vector $\mathbf{w}$, where

$$\mathbf{w} = \mathbf{D}_c^{\frac{1}{2}}\mathbf{x} \qquad (2.16)$$

Rewrite the formula for correlation ratio in terms of this new vector,

$$\eta^2 = \frac{\mathbf{w}'\mathbf{D}_c^{-\frac{1}{2}}\mathbf{F}'\mathbf{D}_r^{-1}\mathbf{F}\mathbf{D}_c^{-\frac{1}{2}}\mathbf{w}}{\mathbf{w}'\mathbf{w}} = \frac{\mathbf{w}'\mathbf{B}'\mathbf{B}\mathbf{w}}{\mathbf{w}'\mathbf{w}} \qquad (2.17)$$

where

$$\mathbf{B} = \mathbf{D}_r^{-\frac{1}{2}} \mathbf{F}' \mathbf{D}_c^{-\frac{1}{2}} \tag{2.18}$$

Therefore the standard form of the eigenequation is given by

$$[\mathbf{B}'\mathbf{B} - \lambda \mathbf{I}]\mathbf{w} = \mathbf{0} \tag{2.19}$$

Once $\mathbf{w}$ is obtained, $\mathbf{x}$ can be obtained by

$$\mathbf{x} = \mathbf{D}_c^{-\frac{1}{2}} \mathbf{w} \tag{2.20}$$

The trivial solution mentioned earlier consists of $\lambda_0 = 1$ and $\mathbf{w}_0 \mathbf{1}$, irrespective of the matrix $\mathbf{B}$, or the data matrix $\mathbf{F}$. To eliminate this trivial solution, we may follow the standard procedure to calculate the residual matrix, say $\mathbf{C}$, by

$$\mathbf{C} = \mathbf{B}'\mathbf{B} - \lambda_0 \frac{\mathbf{w}_0 \mathbf{w}_0'}{\mathbf{w}_0' \mathbf{w}_0} = \mathbf{B}'\mathbf{B} - \frac{\mathbf{D}_c^{\frac{1}{2}} \mathbf{1} \mathbf{1}' \mathbf{D}_c^{\frac{1}{2}}}{f_t} \tag{2.21}$$

In our numerical example, the residual matrix $\mathbf{C}$ is given as follows:

$$\mathbf{C} = \begin{bmatrix} 0.1552 & 0.0006 & -0.1742 \\ 0.0006 & 0.0207 & -0.0250 \\ -0.1742 & -0.0250 & 0.2241 \end{bmatrix}$$

The first component is the solution of the following equation associated with the maximal eigenvalue, say η_1^2,

$$(\mathbf{C} - \eta^2 \mathbf{I})\mathbf{w} = \mathbf{0} \tag{2.22}$$

Once the eigenvector, associated with the largest eigenvalue η_1^2, that is, $\mathbf{w}_1$, is calculated, we obtain the first optimal weight vector $\mathbf{x}_1$ from $\mathbf{w}_1$ by

$$\mathbf{x}_1 = \mathbf{D}_c^{-\frac{1}{2}} \mathbf{w} \tag{2.23}$$

Recall that the weights for the categories are scaled in such a way that

$$\mathbf{w}_1' \mathbf{w}_1 = \mathbf{x}_1' \mathbf{D}_c \mathbf{x}_1 = f_t. \tag{2.24}$$

The corresponding optimal scores for Subject i of Component 1 is given by

$$y_{i1} = \frac{1}{\eta_1} \frac{\sum_{j=1}^{J} f_{ij} x_{j1}}{f_{i.}} \tag{2.25}$$

The vector of scores of the first component is given by

$$\mathbf{y_1} = \frac{1}{\eta_1}\mathbf{D_r^{-1}Fx} \qquad (2.26)$$

Similarly we can express SS_t, SS_b and SS_w in terms of the weights for the rows, that is, $\mathbf{y}$, and maximize the correlation ratio, SS_b/SS_t, subject to the condition that the sum of the weighted responses is zero and that the sum of squares of the weighted responses is equal to f_t. It results in the eigenequation

$$[\mathbf{BB'} - \lambda\mathbf{I}]\mathbf{v} = \mathbf{0} \qquad (2.27)$$

where

$$\mathbf{y} = \mathbf{D_r}^{-\frac{1}{2}}\mathbf{v} \qquad (2.28)$$

2.4 Bivariate Correlation Approach

Recall that our scaling problem is to derive weights for rows and weights for columns. Thus, the response in cell (i,j) is given two weights, y_i and x_j. For f_t responses, there are f_t pairs of weights (y_i, x_j). In our example of Table 2.4, we can express the data in terms of pairs of weights for each response (Table 2.8). Since the two weights in each pair are supposed to describe the same response, it is reasonable to expect that the two weights given to the same response should be as similar as possible. In other words, the correlation defined over those f_t pairs should be as high as possible.

There may be a logical leap here to follow the connection between similar weights in a pair and maximal correlation. Consider an example in which we have two sets (y_i) and (x_i) of five distinct numbers, each consisting of 1, 2, 3, 4 and 5. Then, the sum of the cross products of y_i and x_i is a maximum when $y_i=x_i$. In the current example, $1^2+2^2+3^2+4^2+5^2=55$, which is the maximum, out of all other sums of cross products, $y_i x_i$. The product-moment correlation is proportional to the sum of cross products.

The correlation between responses weighted by y_i and those weighted by x_j, ρ, can be expressed as

$$\rho = \frac{\mathbf{y'Fx}}{\sqrt{\mathbf{x'D_c xy'D_r y}}} \qquad (2.29)$$

Table 2.8 *Doubly Weighted Contingency Table*

	x_1	x_2	x_3
y_1	(y_1, x_1)	(y_1, x_2)	(y_1, x_3)
		(y_1, x_2)	(y_1, x_3)
		(y_1, x_2)	(y_1, x_3)
			(y_1, x_3)
			(y_1, x_3)
			(y_1, x_3)
y_2	(y_2, x_1)	(y_2, x_2)	(y_2, x_3)
	(y_2, x_1)	(y_2, x_2)	(y_2, x_3)
	(y_2, x_1)	(y_2, x_2)	
		(y_2, x_2)	
		(y_2, x_2)	
y_3	(y_3, x_1)	(y_3, x_2)	
	(y_3, x_1)	(y_3, x_2)	
	(y_3, x_1)	(y_3, x_2)	
	(y_3, x_1)		
	(y_3, x_1)		
	(y_3, x_1)		

where

$$\mathbf{f}'_r\mathbf{y} = \mathbf{f}'_c\mathbf{x} = 0 \ \text{ and } \ \mathbf{y}'\mathbf{D}_r\mathbf{y} = \mathbf{x}'\mathbf{D}_c\mathbf{x} = f_t \qquad (2.30)$$

The Lagrangian function for our problem is

$$Q(\mathbf{y}, \mathbf{x},, \lambda_1, \lambda_2) = \mathbf{y}'\mathbf{F}\mathbf{x} - \frac{1}{2}\lambda_1(\mathbf{y}'\mathbf{D}_r\mathbf{y} - f_t)$$
$$-\frac{1}{2}\lambda_2(\mathbf{x}'\mathbf{D}_c\mathbf{x} - f_t) \qquad (2.31)$$

Thus we solve the following sets of equations:

$$\frac{\partial Q}{\partial \mathbf{y}} = \mathbf{0}; \ \ \frac{\partial Q}{\partial \mathbf{x}} = \mathbf{0}; \ \ \frac{\partial Q}{\partial \lambda_1} = 0; \ \ \frac{\partial Q}{\partial \lambda_2} = 0 \qquad (2.32)$$

resulting in

$$\mathbf{F}'\mathbf{y} - \lambda_1 \mathbf{D}_c\mathbf{x} = \mathbf{0}, \quad \mathbf{F}\mathbf{x} - \lambda_2 \mathbf{D}_r\mathbf{y} = \mathbf{0} \tag{2.33}$$

$$\mathbf{x}'\mathbf{D}_c\mathbf{x} - f_t = 0, \quad \mathbf{y}'\mathbf{D}_r\mathbf{y} - f_t = 0 \tag{2.34}$$

From the first two, we obtain the following.

$$\lambda_1 = \frac{\mathbf{x}'\mathbf{F}'\mathbf{y}}{\mathbf{x}'\mathbf{D}_c\mathbf{x}}, \quad \lambda_2 = \frac{\mathbf{y}'\mathbf{F}\mathbf{x}}{\mathbf{y}'\mathbf{D}_r\mathbf{y}} \tag{2.35}$$

Since the denominators of these two expressions are originally set equal, we conclude that the two Lagrange multipliers are equal. Furthermore, we can easily see that Lagrange multipliers are equal to the correlation coefficient, that is, $\lambda_1 = \lambda_2 = \rho$. Considering this and other relations we have so far looked at, we arrive at the conclusion that ρ is equal to the singular value and that $\rho^2 = \eta^2$. Thus, η and ρ, hence ρ^2 and η^2, will be used interchangeably in this book. The first two partial derivatives can be rewritten in the formulas of dual relations (Nishisato, 1980a), or substitution formulas (Benzécri et al., 1973):

$$\mathbf{y} = \frac{1}{\rho}\mathbf{D}_r^{-1}\mathbf{F}\mathbf{x}, \quad \mathbf{x} = \frac{1}{\rho}\mathbf{D}_c^{-1}\mathbf{F}'\mathbf{y} \tag{2.36}$$

This approach, too, results in the same eigenequation as derived for the other approaches. In fact, all these approaches lead to the identical formulation of quantification.

2.5 Geometric Approach

We will use the data in Table 2.4 again. The table contains responses to two categorical variables, teacher and evaluation, each having three categories. The same data can also be represented in the form of (29 judges)-by-(6 categories), the so-called a response-pattern table (Table 2.9).

Each of the two categorical variables (teacher, rating category) has three categories, and each subject's response to each categorical variable is one of the three possible response patterns, (1,0,0), (0,1,0) and (0,0,1), where "1" indicates a choice, the choice being respectively the first, the second and the third category.

These response patterns can be regarded as coordinates of three possible responses, and as such each variable with three categories can be expressed in three-dimensional space with basis coordinates (1,0,0), (0,1,0) and (0,0,1). Each response from a subject falls on one of these

Table 2.9 *Response-Pattern Format of Data in Table 2.4*

S	White	Green	Brown	Good	Average	Poor
1	1	0	0	1	0	0
2	1	0	0	0	1	0
3	1	0	0	0	1	0
4	1	0	0	0	1	0
5	1	0	0	0	0	1
6	1	0	0	0	0	1
7	1	0	0	0	0	1
8	1	0	0	0	0	1
9	1	0	0	0	0	1
10	1	0	0	0	0	1
11	0	1	0	1	0	0
12	0	1	0	1	0	0
13	0	1	0	1	0	0
14	0	1	0	0	1	0
15	0	1	0	0	1	0
16	0	1	0	0	1	0
17	0	1	0	0	1	0
18	0	1	0	0	1	0
19	0	1	0	0	0	1
20	0	1	0	0	0	1
21	0	0	1	1	0	0
22	0	0	1	1	0	0
23	0	0	1	1	0	0
24	0	0	1	1	0	0
25	0	0	1	1	0	0
26	0	0	1	1	0	0
27	0	0	1	0	1	0
28	0	0	1	0	1	0
29	0	0	1	0	1	0

three points, and when we collect responses to a variable from N subjects we have the data (f_1,0,0), (0, f_2,0), (0,0, f_3), where $f_1+f_2+f_3=N$. Then we may ask where these three data points should be located. We can use, for example, the following constraint to determine the locations of the three points:

$$d_1^2 f_1 = d_2^2 f_2 = d_3^2 f_3 = N \qquad (2.37)$$

where d_j is the distance from the origin to category j. Thus, if the frequencies of the three categories are equal to one another, the shape created by connecting the three points is a regular triangle. The above equation means the following: the greater the frequency of the category, the closer the category location to the origin; the smaller the frequency, the further the point from the origin. The coordinates of these three points, thus obtained, are comparable to what we call normed weights or standard coordinates.

Suppose we start with forming a triangle for the three teachers. Then, our task is to project evaluation categories onto this triangle. Similarly, we can form a triangle of the three evaluation categories and then project three teachers onto the evaluation triangle. The coordinates of those projected quantities are projected weights (principal coordinates) on two components. By projection of column categories onto the row space, we mean the following. Consider "Poor" which shows 6 responses for Teacher 1, 2 responses for Teacher 2 and no response for Teacher 3. Thus, the point "Poor" is located on the line connecting Teachers 1 and 2. Divide this line into 8 segments, give 2 segments to Teacher 1 and 6 segments to Teacher 2, which provides the mean value of "Poor," that is, the projection of category "Poor" onto the row (teacher) space. Consider the projection of "Good" onto the teacher space. First, consider the line obtained by connecting Teachers 1 and 2, for which there are 1 and 3 "Good" ratings, respectively. Thus, divide this line into 4 segments, and give 3 segments to Teacher 1 and 1 to Teacher 2. Draw a line from this mean value to Teacher 3. Call this line A. Let us now project "Good" to Teachers 2 and 3, who were rated good 3 and 6 times, respectively, that is, the ratio of 1 to 2. Thus, divide the line from Teachers 2 to Teacher 3 into 3 units, and give 2 to Teacher 2 and 1 to Teacher 3. From this projected point, draw a line to Teacher 1. Call this line B. The projected point of "Good" is where lines A and B cross. Similarly, we can find the projection of category "Average" as shown in Figure 2.3. Figure 2.4 is the projection of Teachers onto the column space of "Evaluation." We can construct the projections of three teachers onto this space as before. Let us note the following points:

(1) the triangles composed from the projected weights are smaller than those of the normed weights. This indicates that the row variable and the column variable are not perfectly correlated. We can speculate then that the closer the projected triangle to the normed triangle, the higher the correlation between the row variable and the column variable. This point will be revisited later.

(2) the projections of the three vertices of the projected triangle onto the horizontal axis has a smaller variance than those of

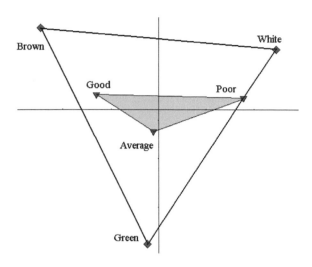

Figure 2.3 *Projection of Evaluation Categories onto Teacher Space*

the normed triangle. When the orientation of the triangle is determined so as to maximize this projection, the ratio of the former to the latter is proportional to the first eigenvalue. This point, too, will be revisited later.

(3) when the triangle orientation is determined, the projection of the three vertices of the projected triangle onto the vertical axis has a smaller variance than those of the normed triangle, and the ratio of the former to the latter is proportional to the second eigenvalue.

(4) the variance of the projections of the three vertices of the normed triangle onto the horizontal axis is the same as that onto the vertical axis. The above geometric analysis of the contingency table is exactly the same with what MUNDA does.

2.6 Other Approaches

As mentioned earlier, there are a number of approaches to arrive at the identical quantification as those discussed so far. Without thorough derivations, let us look at two more approaches.

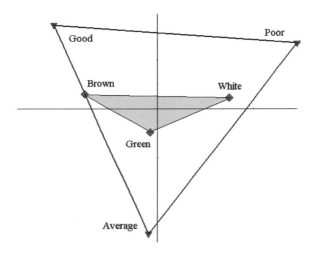

Figure 2.4 *Projection of Teachers onto Evaluation Space*

2.6.1 The Least-Squares Approach

From the ongoing derivations of formulas, we can arrive at a bilinear form of data element f_{ij},

$$f_{ij} = \frac{f_{i.}f_{.j}}{f_t}[1 + \eta_1 y_{i1} x_{j1} + \eta_2 y_{i2} x_{j2} + ... + \eta_K y_{iK} x_{jK}] \qquad (2.38)$$

where we typically assume that

$$\eta_1 \geq \eta_2 \geq \eta_3 \geq ... \geq \eta_k \geq ... \geq \eta_K \qquad (2.39)$$

The first term inside the square bracket corresponds to the trivial solution, which accounts for the portion of data expected when the rows and the columns are statistically independent. Therefore, noting that each component in the bilinear form can be determined as the least-squares approximation to **F**, we can arrive at the least-squares approach, that is, determine weight vectors **y** for rows and **x** for columns of **F** that minimize the squared discrepancy e^2, where

$$e^2 = trace(\mathbf{F} - c\mathbf{y}\mathbf{x}')'(\mathbf{F} - c\mathbf{y}\mathbf{x}') \qquad (2.40)$$

for some constant c. The constant c turns out to be a singular value.

2.6.2 Approach by Cramér's and Tchuproff's Coefficients

Given a two-way table, there are two well-known coefficients, Cramér's coefficient V (Cramér, 1946) and Tchuproff's coefficient of association T (Tchuproff, 1925). First, define the chi square statistic for an $m \times n$ contingency table:

$$\chi^2 = \sum \sum \frac{(f_{ij} - e_{ij})^2}{e_{ij}} \tag{2.41}$$

where

$$e_{ij} = \frac{f_{i.}f_{.j}}{f_t}. \tag{2.42}$$

Pearson(1904) proposed that the correlation between two categorical variables could be expressed as

$$\sqrt{\frac{\phi^2}{1 + \phi^2}} = C \tag{2.43}$$

where

$$\phi^2 = \frac{\chi^2}{f_t} \tag{2.44}$$

where f_t is the total frequency of the $m \times n$ contingency table. One problem with this coefficient was its maximum value of 1 could rarely be approximated in practice unless χ^2 is extremely large. To remedy this problem, Tchuproff (1925) proposed his coefficient T,

$$T = \sqrt{\frac{\chi^2}{f_t \sqrt{(m-1)(n-1)}}} \tag{2.45}$$

T attains its maximum of 1, however, only when $m=n$. Otherwise, its maximum is given by

$$T_{max} = \sqrt{\frac{m-1}{n-1}} \tag{2.46}$$

if $m \leq n$. To remedy this problem, Cramér (1946) proposed his coefficient V

$$V = \sqrt{\frac{\chi^2}{f_t(p-1)}} \tag{2.47}$$

where $p=\min(m,n)$, that is, the smaller of m and n.

    Saporta (1975) discussed orthogonalization of these coefficients which yield coordinates of the variables. In quantification theory, it is well-known that the sum of the eigenvalue associated with the cross-product contingency table is equal to

$$\sum \lambda_j^2 = \frac{\chi^2}{f_t} \tag{2.48}$$

From this relation, we can understand Saporta's consideration of those nonlinear association measures. There are some differences, however, between Saporta's approach and MUNDA approaches so far discussed. Therefore, we will revisit this approach in the last chapter.

2.7 Multidimensional Decomposition

When an eigenequation is involved, it is almost always the case that several components are embedded in the data, and it is necessary to look into multidimensional decomposition of data. Let us define the following matrices:

$$\Lambda = \begin{bmatrix} 1 & 0 & 0 & \dots & 0 \\ 0 & \eta_1^2 & 0 & \dots & 0 \\ 0 & 0 & \eta_2^2 & \dots & 0 \\ . & . & . & \dots & . \\ 0 & 0 & 0 & \dots & \eta_k^2 \end{bmatrix} \tag{2.49}$$

$$\mathbf{W} = [\mathbf{w_0}, \mathbf{w_1}, \mathbf{w_2}, ..., \mathbf{w_k}] \tag{2.50}$$

$$\mathbf{X} = [\mathbf{x_0}, \mathbf{x_1}, \mathbf{x_2}, ..., \mathbf{x_k}] \tag{2.51}$$

where $\mathbf{w_0} = 1$ and $\mathbf{x_0} = \mathbf{D}_c^{\frac{1}{2}} 1$. For each component, the solution of our eigenequation satisfies the following dual relations,

$$y_{ik} = \frac{1}{\eta_k} \frac{\sum_{j=1}^{J} f_{ij} x_{jk}}{f_{i.}} \quad \text{and} \quad x_{jk} = \frac{1}{\eta_k} \frac{\sum_{i=1}^{I} f_{ij} y_{ik}}{f_{.j}} \tag{2.52}$$

It also satisfies the bilinear expression mentioned earlier. For the entire set of solutions, we obtain the following expressions,

$$\mathbf{B'BW} = \mathbf{W\Lambda}, \quad \text{or} \quad \mathbf{W'B'BW} = \mathbf{\Lambda} \tag{2.53}$$

We also note that

$$\mathbf{W'W} = \mathbf{X'D_cX} = \mathbf{I} \tag{2.54}$$

where $\mathbf{I}$ is the identity matrix.

The bilinear form can also be expressed as

$$\mathbf{F} = \frac{1}{f_t} \mathbf{D_r} \mathbf{Y}' \mathbf{\Lambda}^{\frac{1}{2}} \mathbf{X} \mathbf{D_c} \qquad (2.55)$$

where these terms in our numerical example are:

$$\mathbf{F} = \begin{bmatrix} 1 & 3 & 6 \\ 3 & 5 & 2 \\ 6 & 3 & 0 \end{bmatrix}, \quad f_t = 29, \quad \mathbf{D_r} = \begin{bmatrix} 10 & 0 & 0 \\ 0 & 10 & 0 \\ 0 & 0 & 9 \end{bmatrix}$$

$$\mathbf{Y}' = \begin{bmatrix} 1.0000 & 1.0000 & 1.0000 \\ -1.2322 & 0.1227 & 1.2326 \\ -1.9739 & 4.3269 & -2.5056 \end{bmatrix}$$

$$\mathbf{\Lambda} = \begin{bmatrix} 1.0000 & 0.0000 & 0.0000 \\ 0.0000 & 0.6069 & 0.0000 \\ 0.0000 & 0.0000 & 0.1780 \end{bmatrix}$$

$$\mathbf{X} = \begin{bmatrix} 1.0000 & 1.0761 & -0.8608 \\ 1.0000 & 0.0920 & 1.2755 \\ 1.0000 & -1.4126 & -0.6795 \end{bmatrix}, \quad \mathbf{D_c} = \begin{bmatrix} 10 & 0 & 0 \\ 0 & 11 & 0 \\ 0 & 0 & 8 \end{bmatrix}$$

The first component is called trivial solution, and in practice it is deleted from the solution set. When we approximate the data matrix in terms of the trivial and the first k components, the approximation is called the rank k approximation.

In solving the eigenequation

$$(\mathbf{A} - \lambda \mathbf{I})\mathbf{u} = \mathbf{0} \qquad (2.56)$$

successively, where in the current case $\mathbf{A} = \mathbf{BB}'$ or $\mathbf{B}'\mathbf{B}$, we use the formula to calculate the residual matrix, say $\mathbf{C}_1$, when we eliminate $\mathbf{u}_1$ and λ_1 from $\mathbf{A}$,

$$\mathbf{C}_1 = \mathbf{A} - \lambda_1 \frac{\mathbf{u}_1 \mathbf{u}_1'}{\mathbf{u}_1' \mathbf{u}_1} \qquad (2.57)$$

or, more generally,

$$\mathbf{C}_{k+1} = \mathbf{C_k} - \lambda_1 \frac{\mathbf{u_k} \mathbf{u_k}'}{\mathbf{u}_k' \mathbf{u}_k} \qquad (2.58)$$

and

$$\mathbf{A} = \sum_{k=1}^{K} \mathbf{C}_k \qquad (2.59)$$

where $K = rank(\mathbf{A})$.

We have so far discussed a number of equations, and all of them are variations of singular value decomposition (SVD) of a data matrix (e.g., Beltrami, 1873; Jordan, 1874; Schmidt, 1907; Eckart and Young, 1936). SVD is an optimal method of data decomposition as used in principal component analysis, and we can understand from the above equations the reason why Torgerson (1958) called the quantification method 'principal component analysis of categorical data.'

CHAPTER 3

Historical Overview

Simultaneous quantification of rows and columns of a two-way data table has a long history. When we look at it, we cannot help but wonder why a number of researchers have developed essentially the same method under different names, often being unaware of other relevant studies. A reason for this peculiarity perhaps lies first in the fact that there are many ways to formulate the method and second in the fact that the problem of quantification has attracted the attention of researchers in diverse fields of research in many countries. As for the first fact, Greenacre (2005) discussed thirteen ways to formulate "correspondence analysis," one of many names of MUNDA. As for the second fact, it must strictly be due to the point that categorical data are ubiquitous in many fields and many countries.

Although it is not certain when the idea of quantification began, what resembles to MUNDA seems to have originated in the early years of the twentieth century from inquiries by ecologists, biologists, statisticians and psychologists. As we will see shortly, contributions by ecologists and biologists in many countries are overwhelming, and an appropriate credit to them is long overdue.

In 1976, a symposium on optimal scaling, coined by Bock (1960), was held in Murray Hill, New Jersey, USA, with the organizer Forrest Young (USA, Chair), speakers Gilbert Saporta (France), Jan de Leeuw (The Netherlands) and Shizuhiko Nishisato (Canada), and the discussant Joseph Kruskal (USA). It was a land-mark gathering to look at various developments of MUNDA in different countries up to 1976. Around that time, it was still possible to identify major contributors to the field. In his handout, de Leeuw gave a list of the names of those "who have contributed substantially to the data analytic aspect of the problem.

- Originators: Fisher (1935-1940), Guttman (1940-1960), Burt (1950).
- Interesting Discussions, Applications, Specializations: Mosteller (1949), Johnson (1950),Lord (1958), Bock (1960).

- Schools: Hayashi (theory of quantification) (1950-), Benzécri (analyse des correspondances) (1960-)
- Recent Systematizations: Lingoes (1963-1964), McDonald (1967-1969), Nishisato (1972-1973), de Leeuw (1973), Saporta (1975)."

An overview by de Leeuw represented the knowledge among those researchers in the groups of optimal scaling, correspondence analysis, homogeneity analysis and dual scaling at that time.

It is already thirty years since then, however, and we can now find a few descriptions of the history of MUNDA (e.g., Nishisato, 1980a; Benzécri, 1982; Gauch, 1982; de Leeuw, 1983; Greenacre, 1984; Tenenhaus and Young, 1985; Gifi, 1990). Except Gauch (1982), most authors of these publications overlooked extensive contributions to this area by international ecologists and biologists, and their contributions even predated the work by Richardson and Kuder (1933), Hirschfeld (1935), Horst (1935), Fisher (1940), Maung (1941a) and Guttman (1941). This aspect will be revisited shortly. Let us now look at some historical background which nurtured the later developments.

3.1 Mathematical Foundations in Early Days

Algebraic eigenvalue theory was pioneered by such well-known mathematicians as Euler, Cauchy, Jacobi, Cayley and Sylvester in the 18th century (e.g., see de Leeuw, 1973). The idea has been effectively used in many procedures of multivariate analysis as it provides an optimal means for orthogonal decomposition of a square matrix. As such it was put into a practical context, for example, in principal component analysis (PCA). The basic idea of PCA was first presented by Pearson in 1901 and thoroughly formulated by Hotelling in 1933.

When the matrix is restricted to a square matrix, its eigenvalue decomposition can also be regarded as the canonical reduction of bilinear forms $x'Ay$ by two orthogonal transformations of x and y. It is the problem of eliminating the cross product term from the quadratic form $x'Ay$. In this context, some mathematicians made notable and concrete contributions, among others, Beltrami (1873), Jordan (1874) and Schmidt (1907). The basic idea was then generalized to any rectangular matrix by Eckart and Young (1936) under the name of the *Eckart-Young decomposition theorem*, which is now more popularly known as singular value decomposition (SVD), as opposed to its square-matrix counterpart, eigenvalue decomposition (EVD).

3.2 Pioneers of MUNDA in the 20th Century

(1) **International Group of Ecologists and Biologists**: In the early part of the 20th century, a number of ecologists were engaged in advancing ecological theory of plant growth patterns and developed an analytical procedure, gradient and ordination method. According to Whittaker (1967), it was Ramensky (1930), a Russian ecologist, who should be regarded as the originator of "gradient analysis," rather than Gleason (1926) in the USA, Lenoble (1927) in France or Ellenberg (1948) in Germany. These ecologists nurtured the idea of weighted averaging scheme, which led to the procedure called "ordination," a concept closely related to the method of reciprocal averages. Thus, it seems that the idea of MUNDA has originated from their work, rather than the work by Fisher, Hirschfeld (Hartley), Guttman, Hayashi or Benzécri. The work by ecologists and biologists was obviously not well known among those who developed a number of procedures relevant to MUNDA in later years. This early independent development is indicative of the history of MUNDA.

When the idea of SVD and EVD is looked at from the MUNDA point of view, we are moving from analysis of continuous variables to that of categorical variables. Then, the idea of quantification of data became a center of activities for some other researchers as well. Let us identify a few of them.

(2) **Richardson and Kuder** (1933): They are better known for the 'Kuder-Richardson reliability coefficient' in test construction than for their pioneering work on quantification. In scoring multiple-choice data, they came up with the idea which was later called by Guttman (1941) the principle of internal consistency. This principle was applied to deriving a single variate from a set of qualitative variates by Horst (1936), and Edgerton and Kolbe (1936). The same principle was used for quantitative variables by Wilks (1938).

In arriving at internally consistent scores for subjects, Richardson and Kuder (1933) used a procedure, which Horst (1935) called the *method of reciprocal averages* (MRA). As we saw in the previous chapter, MRA is indeed an established procedure for MUNDA, but it looks as though the idea was pursued by ecologists before Richardson and Kuder.

(3) **Hirschfeld (H.O. Hartley)** (1935): Even those in applied areas of statistics know his famous book with E. Pearson, entitled *"Biometrika Tables for Statisticians."* To quantification theory, he made a remarkable and outstanding contribution. His 1935 question was "Given a two-way table, is it possible to assign weights to the rows and the

columns in such a way that the regression of rows on columns and the regression of the columns on the rows be simultaneously linear?"

Although his question looks quite different from that of Richardson and Kuder, the idea is very similar to that of MRA. Rather than an iterative process used in MRA, Hirschfeld provided an algebraic solution to his question. Many years later, Lingoes (1964) talked about the *method of simultaneous linear regressions* to describe Guttman's approach (Guttman, 1941). This name, however, seems to be more appropriate to describe Hirschfeld's procedure than Guttman's. Lingoes' association of the name with Guttman's work suggests implicit similarity of Guttman's approach to Hirschfeld's work.

(4) **Fisher** (1940): No need to explain the fame of Fisher for his contributions to statistics. It is interesting to note, however, that the quantification problem also attracted his attention. Given data in the form of the contingency table, Fisher considered discriminant analysis of categorical data. More specifically, his problem was what weights to the columns of a contingency table would maximally discriminate between the rows. Thus, this was the problem to determine weights for the columns of a contingency table so as to maximize the between-rows sum of squares relative to the total sum of squares. It is interesting to note that for the computation he, too, suggested essentially the same process as used by MRA.

(5) **Guttman** (1941) and **Maung** (1941a): Guttman was one of the most influential psychometricians in the twentieth century, and his work laid foundations in scaling, factor analysis and test theory. Maung seems to have been in the shadow of Fisher, in spite of his great contribution to quantification theory. Both Guttman and Maung presented remarkably thorough descriptions of quantification procedures, Guttman on quantifying multiple-choice data and Maung on contingency tables. Guttman explored the principle of internal consistency. Guttman admits that his approach is similar to Richardson and Kuder, Horst, Edgerton and Kolbe, mentioned above, but insists that his theory was motivated by other considerations and was developed independently of those others. He proved that the three approaches derived from it lead to the identical quantification:

• Determine weights for subjects so as to maximize the between-options sum of squares,

• Determine weights for options so as to maximize the between-subjects sum of squares, and

• Determine weights for both subjects and options so as to maximize the product-moment correlation between data weighted for subjects and those for options.

Giving credits to Fisher as Fisher's scoring procedure, Maung elaborated quantification of the contingency table in three distinct ways:

• Determine weights for rows and columns so as to maximize the product-moment correlation between data weighted for the rows and those for the columns,

• Determine weights for the rows so as to maximize the ratio of the between-columns sum of squares to the total sum of squares, and

• Determine weights for the categories so as to maximize the canonical correlation between two sets of categories, one for the rows and the other for the columns.

Again Maung provided that the three approaches yield identical results. He also presented an application of the Fisher-Maung method to real data (Maung, 1941b). Both Guttman and Maung provided very thorough mathematical formulations of their procedures.

(6) **Guttman** (1946): He further extended his approach of internal consistency to different data types, rank order and paired comparisons. This paper should be regarded as Guttman's another monumental work after his 1941 paper. As we will see later, most quantification methods, called under different names, are restricted to the quantification of the contingency table and multiple-choice data. Guttman's 1946 paper extended it to the so-called dominance data, and laid a solid foundation for later studies.

3.3 Rediscovery and Further Developments

In retrospect, the foundations for quantification theory were firmly laid by those researchers mentioned above. Yet, mathematically equivalent procedures for quantification have been proposed, often using different objective functions for optimization and often under new names. This aspect is peculiar to the history of quantification theory, thus noteworthy.

With Guttman's 1946 paper, most aspects of the quantification method were solidly established. However, there have been a number of important studies, for example, implementation of the method of reciprocal averages into the IBM sorting cards by Mosier (1946) and consideration of data, consisting of such codes as −, -, ?, + and ++, for the analysis of variance by Fisher (1948). The readers may also be interested in such publications as Johnson (1950), Hayashi (1950, 1952), Bartlett (1947), Williams (1952), Lancaster (1953, 1958), Bock (1956, 1960), Lord (1958) and Torgerson (1958). After these publications, we see a dramatic increase in publications relevant to MUNDA, and it becomes difficult to trace the history without injustice to some researchers. Nevertheless, let us try to identify some of the readily notable cases.

3.3.1 Distinct Groups

There are many researchers who provided centers of activities. Some of them are:

(A) Gradient-Ordination Research Group

Gauch (1982) states that "substantial ordination work was done in ecology around 1950, using simple algorithm and hand calculation." Following Gleason in the USA, Lenoble in France, Ramensky in Russia and Ellenberg in Germany, this group has produced almost a countless number of publications. Since the current book does not introduce this group's contributions to MUNDA, some major references are provided below. Examples of weighted average ordination and gradient studies are, among others, Whittaker (1948), Curtis and McIntosh (1951), Rowe (1956), Loucks (1962), Whittaker (1966, 1967), Hill (1973), Gauch, Chase and Whittaker (1974), Gauch and Wentworth (1976), Austin (1976), Gauch, Whittaker and Wentworth (1977), Gauch, Whittaker and Singer (1981), Digby and Gower (1981), Oksanen and Ahti (1982), Goodall and Johnson (1982), Heiser (1982), Ihm and Van Groenwoud (1984), Ter Braak (1986, 1987), and Kzranowski (1994).

These are only a handful of references, and the interested readers can start with these and then move into an enormous number of impressive applications of the basic procedures to studies in biology and ecology. A very unique aspect of this international group seems to lie in the fact that their gradient-ordination method and its variants were developed in close relation to ecological and biological work, almost to the extent that an analytical needs of their data preceded the developments of the methods. When we look at researchers in the literature, we are overwhelmed with the abundance of names of ecologists and biologists from many countries, and many non-Anglo Saxon names, indicating a wide spread of research activities. Thus, this is a group heavily oriented towards practically useful methods of MUNDA, and they have been highly successful in their endeavors.

(B) Hayashi School

In Japan, headed by Chikio Hayashi since 1950. The work is known as *Hayashi's theory of quantification*, which started with Hayashi's studies in 1950 and 1952. This group developed a comprehensive scheme of data analysis for different objectives with different procedures. The quantification of the contingency table and multiple-choice data as discussed in this book is only one of them and known as Hayashi's quantification theory Type III. Its detailed history from 1950 to 1970 was traced in two Japanese papers by Morimoto (1997a, 1999) and a brief history in English (Morimoto, 1997b). This group provided a strong guidepost for many statisticians, psychometricians, sociologists, educational researchers, econometricians and marketing researchers in Japan.

(C) Benzécri School

In France, headed by Jean-Paul Benzécri, started in the early 1960s. The work is known as *correspondence analysis* for contingency tables and *multiple correspondence analysis* for multiple-choice data. The journal "Les Cahiers de l'Analyse des Données" played a major role in dissemination of French correspondence analysis. The journal, however, ceased to exist in 1996. This group emphasized the geometric aspects of quantification, geared primarily towards analysis of contingency tables and multiple-choice data. A monumental book by Benzécri and his many collaborators was published in 1973. The French group dominated the world of quantification in the 1970s and the 1980s.

(D) Leiden Group

In The Netherlands, headed by Jan de Leeuw, Willem Heiser, Jacqueline Meulman and others, started in the late 1960s. This group developed the Gifi system of quantification, often referred to as break homogeneity analysis. The work involved generalizing the loss function, imposing restrictions for quantification and incorporating other popular multivariate methods into the system. This group made Leiden as a world center of activities in psychometrics within a short period of some thirty years. An outstanding book on the Gifi system was published in 1990 (Gifi, 1990), with the collaborations of de Leeuw, van de Geer, Heiser, Meulman, van der Burg, van Rijckevorsel, van der Heijden, Bettonvil, Stoop, Meester, van den Berg, Groenen, Verboon, Verdegaal, de Bie, Knip, and Nierop. Excellent compendia are also available (de Leeuw, 1984a; Michailidis and de Leeuw, 1998).

(E) Toronto Group

In Canada, headed by Shizuhiko Nishisato, started in the late 1960s. This group identified two types of categorical data, incidence data and dominance data, fundamentally different from the quantification point of view, in an attempt to quantify a wider variety of data. It was successful in building a comprehensive framework for analysis of contingency tables, multiple-choice (response-pattern) data, sorting data, rank-order data, paired comparison data, successive categories data and multi-way data. The work is known under the name dual scaling, and two books by Nishisato (1980a, 1994) summarize developments of dual scaling, followed by the current book on MUNDA.

(F) Chapel Hill Group

In the 1970s, three eminent psychometricians, Forrest Young, Jan de Leeuw and Yoshio Takane, worked together at the L.L.Thurstone Psychometric Laboratory, University of North Carolina at Chapel Hill, and published a series of papers in *Psychometrika*. One of them (de Leeuw, Young and Takane, 1976) is particularly important for MUNDA because they expanded the basic approach to any levels of measurement (nominal, ordinal, interval and ratio) for a number of objectives, called processes. It was a comprehensive framework for quantification. Young was further involved in an attempt to synthesize quantification methods and visual display (Young, 1981; Tenenhaus and Young, 1984), de Leeuw continued his work at Leiden University and the University of California at Los Angeles, and Takane moved to McGill University, where he has made an outstanding contribution to psychometrics and linear algebra. For marketing researchers, Perreault and Young (1980) and Hoffman and Franke (1986) are recommended. We must not forget important contributions by Bock (1956, 1960) and McKeon (1966) in early days.

(G) Biplot Group

In the USA (Gabriel, Odoroff), Europe (Aitchison, Carlier, Cox, Gower, Greenacre, Hand, Harding, Israëls, Kroonenberg, Sikkel, Ter Braak, Underhill), and Asia (Han and Huh), who carved a new insight into graphical display of quantification. In particular, Gower is instrumental to its generalizations to multidimensional space and to nonlinear biplots (Gower, 1990, 1992; Gower and Harding, 1988). Their rigorous treatment of graphical research into MUNDA is outstanding, and the work was culminated into a book by Gower and Hand (1996). The key players in this group have also published many outstanding papers, relevant to MUNDA.

(H) Association-Model and Correlation-Model Group

In development of models for loglinear analysis, research was extended to correspondence analysis. Some of the major publications are Daudin and Trecourt (1980), Lauro and Decarli (1982), Fienberg and Meyer (1983), van der Heijden and de Leeuw (1985), Leclerc, Chevalier, Luce and Blanc (1985), Goodman (1985a, 1986, 1991), Escoufier and Junca (1986), Gilula and Haberman (1986), Choulakian (1988), van der Heijden, de Falguerolles and de Leeuw (1989), Whittaker (1989), and Gilula and Ritov (1990). The Ph.D. thesis by Millones (1991) offers a well-written summary of comparisons between association models, partially optimal scaling (Inukai, 1971; Nishisato and Inukai, 1972) and dual scaling (Nishisato, 1980, 1994).

(I) Non-Symmetric Correspondence Analysis Group

A new group of statisticians in Italy launched an investigation to what they call non-symmetric correspondence analysis in 1983 by Lauro and D'Ambra, and since then they and other researchers have been working on its extensions to three-way tables and ordered categories. Some of the references are Escoufier and Grorud (1980), Lauro and D'Ambra (1984), Goodman (1985b), Lauro and Siciliano (1988), D'Ambra and Lauro (1989), Siciliano, Mooijaart and van der Heijden (1990), D'Ambra and Lauro (1991), Kroonenberg and Lombardo (1998, 1999), Greenacre (2000), D'Ambra, Lombardo and Amenta (2002), D'Ambra, Beh and Amenta (2005), and Kroonenberg (2002). Many of these researchers are extending their work towards ordered categories, using orthogonal polynomials (e.g., Beh, 2001b), a newly emerging approach.

The topic of asymmetric data matrices has also been investigated extensively by people outside MUNDA, for example, see excellent contributions by such researchers in multidimensional scaling as Okada, Imaizumi, Chino and their colleagues (e.g., Okada and Imaizumi, 1997, 2000, 2002, 2003, 2005; Okada, Imaizumi and Inoue, 2005; Chino, 1978, 1990, 2002; Chino and Shiraiwa, 1993).

(J) Sensitivity Analysis Group

Mainly in Japan, a group of statisticians contributed to sensitivity analysis of MUNDA. As we are all aware, MUNDA being descriptive, we are constantly concerned with the stability of scaling results. Tanaka (1978) worked out asymptotic theories of MUNDA, and he and his group provided a means to assess stability of results. Since this topic is not discussed in detail in the current book, we provide relevant publications.

As is well-known, sensitivity analysis is very popular and important in multivariate analysis. The following list, however, does not include an enormous number of papers on sensitivity analysis for principal component analysis, factor analysis and regression analysis, but only those relevant to quantification of categorical data. They are Tanaka and Kodake (1980), Tanaka (1983, 1984a,b, 1992), Tanaka and Tarumi (1985, 1988a,b,c), Tarumi (1986), Tarumi and Tanaka (1986) and Huh (1989).

In addition, we should mention a one-man team Roderick P. McDonald who made a substantial impact on quantification research through his publications in a wide area of psychometrics (e.g., McDonald, 1968, 1983; McDonald, Torii and Nishisato, 1979). He was also a mentor to the Toronto group for ten years.

3.3.2 Books and Papers

One of the effective means of dissemination and popularization is through publications of books. The work on quantification theory started in Japan in the 1950s and France in the 1960s, and in these countries their methods were put into practice, particularly effective for governmental work. A number of books on quantification were published in the 1960s, 70s and 80s in France and Japan.

In French, we have such books as Lebart and Fénelon (1971), Benzécri et al. (1973), Carliez and Pagés (1976), Escofier and Le Roux, (1976); Lebart, Morineau and Tabard (1977), Bouroche (1977), Jambu and Lebeaux, (1978), Saporta (1979), Lebart, Morineau and Fénelon (1979), Benzécri and Benzécri (1980), Fénelon (1981), Foucart (1982, 1985), Nakache (1982), Cibois (1983), Escofier and Pagés (1988), Jambu (1989), Saporta (1990), Rouanet and Le Roux (1993), Tenenhaus (1994), Celeux and Nakache (1994), and Escofier (2003).

In Japanese, we have such books as Hayashi, Higuchi and Komazawa (1970), Komazawa (1978, 1982), Takeuchi and Yanai (1972), Hayashi (1974, 1992), Nishisato (1975, 1982), Saito (1980), Kobayashi (1981), Hayashi and Suzuki (1986), Takane (1980a), Iwatsubo (1987), Akiyama (1993), Ohsumi, Lebart, Morineau, Warwick and Baba (1994), and Komazawa, Hashiguchi and Ishizaki (1998).

An irony of the matter is that many researchers in English-speaking countries were more or less oblivious of French and Japanese work half a century ago, to the extent that Hill (1974) called quantification theory "neglected multivariate analysis."

In ecology and biology, however, there were books in English on "gradation and ordination" methods such as Curtis (1959), Everitt (1978), Orlóci (1978), Whittaker (1978b), Gauch (1982) and Legendre and Legendre (1994).

In spite of their close relation to MUNDA, an enormous amount of contributions by ecologists and biologists somehow mostly escaped the attention of researchers outside the disciplines. Outside ecology and biology, the English-speaking community did not seem to have seen those books and earlier papers by ecologists and biologists. Only after 1980s, we see books on MUNDA in English, such as Nishisato (1980a, 1994), Greenacre (1984,1993a), Lebart, Morineau and Warwick (1984), Nishisato and Nishisato (1984, 1994), van Rijckevorsel and de Leeuw (1988), Gifi (1990), Weller and Romney (1990), van de Geer (1993), Greenacre and Blasius (1994, 2006), Gower and Hand (1996), Blasius and Greenacre (1998), Clausen (1998), Benzécri (1992), Le Roux and Rouanet (2004) and Murtagh (2005).

In early days, many researchers in English-speaking countries were not familiar with French or Japanese contributions. For instance, Nishisato (1980a) lists only a few French studies, namely those by Benzécri (1969), Benzécri and others (1973), Escofier-Cordier (1969), Saporta (1975), and Bouroche, Saporta and Tenenhaus (1975). His attention to these studies was drawn by a letter from Michael Browne (June 17, 1976), then in South Africa, who stated that a method equivalent to optimal scaling (the name proposed by R. Darrell Bock (1960), and used in Nishisato's earlier studies) was popular in France, and that the researchers there would plot rows and columns jointly in the same space. Browne's view that the mathematical justification of plotting rows and columns together was not sound deeply affected Nishisato (1980a) who consequently refrained from using graphical displays in the book.

The language barrier between the English-speaking and the French-speaking researchers was remedied partly by the symposium on optimal scaling at the 1976 annual meeting of the Psychometric Society, organized by Forrest W. Young with speakers de Leeuw (The Netherlands), Saporta (France) and Nishisato (Canada), and partly by the Workshop on Nonmetric Data Analysis in Paris, organized by Michel Tenenhaus (France), where Benzécri delivered a special lecture on French correspondence analysis. It was a pre-session for the Third International Meeting of the Psychometric Society, at Joux-en-Josas in 1988. The language barrier, however, existed for a long time between the English-speaking community and the rest of the world.

Greenacre's 1984 book played an important role in improved communication between the French-speaking and the English-speaking researchers. Greenacre, trained under the guidance of Benzécri, has also given talks on correspondence analysis throughout the world. In much lesser degrees, Nishisato's two books in Japanese (Nishisato, 1975, 1982) provided new pieces of information to Japanese researchers (Iwatsubo, 1975; Yamada, 1994). Integrations of different language groups were gradually promoted through international conferences, in particular those held in France (several volumes of "Analyse de données et informatique" (Data analysis and informatics)), annual meetings of the German Classification Society, meetings of the International Federation of Classification Societies, meetings of the Classification Society of North America, annual meetings of the Psychometric Society, and those international meetings on quantification methods, organized by Greenacre and Blasius. It should also be noted that ecologist Ter Braak actively contributed to many psychometric meetings and acted like an ambassador of the ecology group to the psychometric and statistic communities, and that wide publications by Everitt helped filling the gap between different disciplines.

In the midst of international communication, a remarkable group of talents in psychometrics emerged, that is, the Leiden group. Leiden University graduates published a number of top-quality doctoral theses and additional work in English from DSWO Press, to name a few, de Leeuw (1973, 1984b), Heiser (1981), Meulman (1982, 1986), van der Heijden (1987), van Rijckevorsel (1987), Israëls (1987), van der Burg (1988), Koster (1989), Kiers (1989), van Buuren (1990), Verboon (1994), Markus (1994) and van Os (2000). These theses made noteworthy contributions to the dissemination of quantification theory, not only among English-speaking countries but also throughout the rest of the world.

In the 1970s and onward, there were many other individual researchers, who contributed to the developments of quantification theory. Out of many publications, there are several papers which offer excellent overviews of quantification: de Leeuw Young, and Takane (1976), as mentioned earlier; Tenenhaus (1982), who presented a synthesis of variants of correspondence analysis in terms of duality scheme; Bock (1982) for German speaking researchers; Tenenhaus and Young (1985) who presented a synthesis of quantification methods under different names; four presidential addresses of the Psychometric Society, all published in *Psychometrika*, namely Young (1981) on outlining the boundaries of quantification problems, Nishisato (1996) on clarifying many hidden aspects of dual scaling, Meulman (2003) on re-focusing quantification problems on distance analysis and Heiser (2004) on geometric formulations of quantification; an excellent, detailed and helpful comparison of several formulations of quantification problems by Takane, Yanai and Mayekawa (1991); Michailidis and de Leeuw (1998) on the Gifi system of descriptive multivariate analysis.

Many new ideas also emerged, among others, Rao (1995a,b) discussed some disadvantages of using the chi-square metric distance, most notably that it gives very large weights to categories with very low frequencies, and proposed an alternative metric, that is, the Hellinger distance. Given two sets of relative proportions

$$p_i = (p_{i1}, p_{i2}, \cdots, p_{im}), p_j = (p_{j1}, p_{j2}, \cdots, p_{jm}) \qquad (3.1)$$

the Hellinger distance is given by

$$D_H(i,j) = \sqrt{\sum_{k=1}^{m} (\sqrt{p_{ik}} - \sqrt{p_{jk}})^2} \qquad (3.2)$$

The use of the Hellinger distance mitigates the effect of the sample sizes, and it is dependent only on the concerned pair of variables, rather than on all the variables. Rao presented numerical examples. An application of the Hellinger distance can also be found in Cuadras and Fortina (1998), where they state that this distance measure is a sensible choice when the frequency table has a multinomial structure. Cuadras is a key researcher who published many papers on graphical display of categorical data, too many to list all the relevant publications by him and his colleagues here. Nakayama Naito and Fujikoshi (1998) discussed the stability issue of correspondence analysis and analysis using the Hellinger distance as its alternative. One possible drawback of the chi-square distance is that the weight of a category with a small frequency becomes exceptionally large. In this regard alone, Rao's proposal is worth pursuing and opened a new horizon.

There are other interesting studies, to name only a few, the use of a different norm by Choulakian (2005) in his taxi cab correspondence analysis, another norm by Yamakawa, Ichihashi and Miyoshi (1998), and an optimization scheme, called regularization and used for MUNDA, by Takane and Hwang (2005). The Hellinger distance, different norms and regularization methods are all suggestive of new directions in MUNDA's development.

3.3.3 A Plethora of Aliases

One of the distinct features of MUNDA is that many researchers in diverse areas of specialty in several countries have proposed different names for essentially the same technique. Some of the widely known such names are gradient analysis, ordination, reciprocal averaging, simultaneous linear regressions, Fisher's appropriate scoring, additive scoring, Hayashi's theory of quantification, Bock's optimal scaling, correspondence analysis, homogeneity analysis and dual scaling. If we include in this list variants of these names and procedures, and arrange all of them in chronological order, we can see a vivid line of thoughts in the developments of MUNDA, towards special conditions, constraints, potentials and purposes. Relatively popular names of MUNDA and its variants are in chronological order:

- Gradient analysis (Ramensky, 1930; Whittaker, 1948)
- The method of reciprocal averages (Richardson and Kuder, 1933; Horst, 1935)
- Simultaneous linear regression (Hirschfeld, 1935; Lingoes, 1964)
- Appropriate scoring and additive scoring (Fisher, 1948)
- Hayashi's theory of quantification (Hayashi, 1950)
- Principal component analysis of categorical data (Torgerson, 1958)
- Optimal scaling (Bock, 1960)
- Ordination (Whittaker (1967) and others)
- Analyse des correspondances (Benzécri, 1969; Escofier-Cordier, 1969)
- Biplot (Gabriel, 1971)
- ANOVA of categorical data (Nishisato, 1971)
- Partially optimal scaling (Inukai, 1972; Nishisato and Inukai, 1972)
- Categorical conjoint measurement (Carroll, 1973)
- Homogeneity analysis (de Leeuw, 1973)
- Reciprocal averaging (Hill, 1973)
- Correspondence factor analysis (Teil, 1975)
- Non-metric ordination method (Prentice, 1977)
- Direct gradient analysis (Whittaker, 1978a)
- Basic content scaling (Jackson and Helmes, 1979)
- Dual scaling (Nishisato, 1980a)
- Detrended correspondence analysis (Hill and Gauch, 1980)
- Nonparametric ordination (Gauch, Whittaker and Singer, 1981)
- Centroid scaling (Noma, 1982)
- Nonlinear ordination (Goodall and Johnson, 1982)
- Joint ordination (Heiser, 1982)
- Multivariate descriptive statistical analysis (Lebart, Morineau and Warwick, 1984)
- Forced classification (Nishisato, 1984a)
- Non-symmetric correspondence analysis (Lauro and D'Ambra, 1984; D'Ambra and Lauro, 1989; Kroonenberg and Lombardo, 1999)

- Residual ordination analysis (Carleton 1984)
- Canonical correspondence analysis (Ter Braak, 1985, 1986)
- CGS scaling (Carroll, Green and Schaffer, 1986)
- Partial correspondence analysis (Yanai, 1987, 1988)
- Joint correspondence analysis (Greenacre, 1988, 1994)
- Partial canonical correspondence analysis (Ter Braak, 1988)
- Nonsymmetrical three-way correspondence analysis (D'Ambra and Lauro, 1989)
- Nonsymmetrical partial correspondence analysis (D'Ambra and Lauro, 1989)
- Canonical analysis with linear constraints (Böckenholt and Böckenholt, 1990)
- Nonlinear multivariate analysis (Gifi, 1990)
- Inferential ordinal correspondence analysis (Gilula and Ritov, 1990)
- Three-way biplot (Gower, 1990)
- Standardized dual scaling (Nishisato 1991)
- Generalized biplot (Gower, 1992)
- Three-way correspondence analysis (Carlier and Kroonenberg, 1995, 1996)
- Descriptive multivariate analysis (Michailidis and de Leeuw, 1998)
- Constrained correspondence analysis (Groenen and Poblome, 2002)
- Generalized constrained multiple correspondence analysis (Hwang and Takane, 2002)
- Partial multiple correspondence analysis (Yanai and Maeda, 2002)
- Multiple correspondence spline analysis (Adachi, 2004a)
- Inverse correspondence analysis (Groenen and van de Velden, 2004)
- Taxicab correspondence analysis (Choulakian, 2005)
- Regularized multiple correspondence analysis (Takane and Hwang, 2005)
- Multidimensional nonlinear descriptive analysis (Nishisato, 2006)

In early days, different names meant typically the same mathematical procedure, that is, singular value decomposition. Gradually, however, many variants of MUNDA were proposed, which were not always based on direct applications of singular value decomposition, but on modified or more constrained frameworks than the original formulation. All these names, therefore, indicate historical developments of MUNDA's methodology.

3.3.4 Notes on Dual Scaling

Since dual scaling is what the current author coined and MUNDA is more or less a compendium of what has been done in dual scaling, a short note on the name may be of some use. In 1976, a symposium on optimal scaling was held in Murray Hill, New Jersey, with Young, Saporta, de Leeuw, Nishisato and Kruskal (USA) as participants. In the discussion session, Zinnes (USA) raised the question if the popular name "Optimal Scaling," was appropriate, or rather specific enough for the procedure. Nishisato then proposed a new name "Dual Scaling" as an alternative for the reason that the method is based on symmetric bilinear decomposition. With the generally positive support for the name by the audience at the symposium, he adopted it in his 1980 book, which was perhaps the first comprehensive book on MUNDA in English. In spite of this historical background, however, the name dual scaling has not always been preferred to such a popular name as correspondence analysis. As a minority expression, Franke (1985) states that he uses "Nishisato's term for its generality and lack of ambiguity"(p.63).

Until recently, dual scaling is a distinct procedure because of its applicability to a wider range of categorical data than others. Nishisato has extended its applicability to a wide range of data identifying two distinct objectives in scaling, one for incidence and the other for dominance data. This distinction also coincided with two distinct distance measures (chi-square metric for incidence data and Euclidean metric for dominance data). This general aspect of dual scaling led to the statement that dual scaling is "a comprehensive framework for multi-dimensional analysis of categorical data" (Meulman, 1998). Michialidis and de Leeuw (1998) also states that "...most of them have passed the stage of basic formulation and moved toward their own unique advancement. Hence, we have Nishisato's efforts to apply dual scaling technique to a wider variety of data ..." Recently, however, researchers favoring the name correspondence analysis have extended the method to dominance data (e.g., Greenacre and Torres-Lacomba, 1999; Beh, 1999; van de Velden, 2000; Torres-Lacomba and Greenacre, 2002). Thus, the difference between dual scaling and correspondence analysis has finally diminished, except for some minor peculiarities of formulations and conceptual stances.

It is the author's view, however, that MUNDA has not yet reached the peak of maturity as a data analytic method, the most lacking being a rigorous treatment of the joint row-column association structure: as this will be further discussed in 5.5.4, we know that the space for row weights and that for column weights are different, and that joint graphical display, therefore, requires special caution; yet we have never developed an appropriate procedure for joint multidimensional decomposition of both within-set (i.e., between rows and between-columns) distances and between-set (i.e., between a row and a column) distances, a problem presented by Nishisato and Clavel (2002). Some other problems for the future direction are discussed in the last chapter, entitled "Further Perspectives."

3.4 Additional Notes

At the last international conference on Correspondence Analysis and Related Methods, organized by Michael Greenacre and Jörge Blasius and held at Univesidad Pompeu Fabra, Barcelona, in 2004, there were many participants from Spain, in addition to invited speakers from such countries as France, Great Britain, The Netherlands, Italy, Japan, Canada and Spain. It was an eye-opening event to see how widely MUNDA has been studied in Spain. The conference was dedicated to the truly outstanding contributor who passed away recently, Ruben Gabriel, known for his biplot. At the conference, Michael Greenacre also mentioned the passing of Chikio Hayashi of Hayashi's theory of quantification and Brigette Escofier, a key researcher in French correspondence analysis. At the International Meeting of the Psychometric Society, held in Tilburg, The Netherlands, in 2005, there were symposia and sessions on MUNDA-related topics, indicating that research on MUNDA was still going and strongly alive. Those interested in the relevant literature may find the following bibliographic publications very useful: Nishisato (1986c), Birks, Peglar and Austin (1994, 1996) and Beh (2004).

3.4.1 Dedications

MUNDA is 80 years old if we start with the paper by Gleason (1926), 76 years from Ramensky (1930), or 73 years from Richardson and Kuder (1933). MUNDA is an unusual topic in the sense that it has attracted researchers in so many disciplines and so many countries. To pay the respects to those researchers who contributed to its history, it was planned to compile a list of contributors to this field. The task, however, soon met difficulties with almost endless encounters with relevant studies, particularly those applied papers of MUNDA-related techniques in ecology and biology. This task also made me realize how inadequate my knowledge of the area was, utterly inappropriate to write a book on the entire spectrum of the relevant topics.

A list of the contributors was initially prepared with the kind help of Yasumasa Baba (Japan), Jörge Blasius (Germany), John Gower (Great Britain), Ludovic Lebart (France) and Yoshio Takane (Canada). It was a very time-consuming task, and when the list exceeded 3500 names, however, the original plan of putting it in the appendix of this book had to be abandoned.

It was a good exercise, though, to see many foreign and often unusual names of researchers, indicating how international this topic of research has been. The current list, mentioned above, will be further augmented and will hopefully be posted on my website at a later date. In this book, only a very small fraction of the names from that list are cited, but we should remember that there are many other hidden researchers whose studies have provided a healthy ingredient for the development of MUNDA. My sincere salute to all those contributors!

CHAPTER 4

Conceptual Preliminaries

4.1 Stevens' Four Levels of Measurement

Measurement is typically defined as an assignment of numerals to objects according to certain rules. Stevens (1951) classified measurement into four types: nominal, ordinal, interval and ratio.

Nominal measurement. An example of nominal measurement is the numbering of athletes, such as 3, 5, 7 and 10, where these numbers are used for identification or as labels. The only mathematical rule to govern nominal measurement is one-to-one correspondence. If two objects have different numbers or labels, they are considered different.

Ordinal measurement. An example of ordinal measurement is the ranking of movies, such as 1, 2 and 3, where 1 indicates the best, 2 the second best and 3 the last one, without providing any information how much the best is better than the second best. Here the mathematical rules are one-to-one correspondence and order relations for which only a monotone transformation is permissible.

Interval measurement. An example of interval measurement is temperature in Celsius or Fahrenheit, where the difference is meaningful (e.g., the difference between 3 and 5 degrees is considered the same as the difference between 10 and 12 degrees). However, the measurement does not have the rational origin and thus cannot be used in such a way that the temperature of 20 degrees is twice as hot as 10 degrees. Imagine what happens if we change the temperature from Celsius to Fahrenheit. Here the mathematical rules are one-to one correspondence, ranking and equality of the unit.

Ratio measurement. An example of ratio measurement is distance, say between Toronto and Montreal, where the measurement has the absolute origin (i.e., "0" means nothingness of the attribute under consideration). The existence of the origin allows the ratio to be a meaningful quantity. For instance, we can now say that the distance of 10 km is twice the distance of 5 km. Here the permissible mathematical operations are one-to-one correspondence, ranking, equality of the unit and the division and multiplication.

In psychometrics, there is a major branch called scaling, of which objective is to upgrade the measurement level. If the input data are nominal, the object of scaling is to transform the data to ordinal, interval or ratio level. Data types often handled by scaling, therefore, are nominal and ordinal measurement, which together are usually called qualitative data or categorical data. When data are at the level of interval or ratio measurement, there is not much work needed to further upgrade them by scaling.

For traditional inferential statistical methods, we typically regard data as a random sample from the population with the normal distribution. This setup, however, has great demerits as well as merits. It enables the researchers to take advantage of the normal distribution in the assessment of the likelihood of an outcome under a specified condition, and at the same time this setup limits us to consider only linear correlation between variables, for the bivariate normal distribution does not contain any other relationship than linear. In data analysis, this restriction may become crucial. Furthermore, if the normal distribution is assumed, the level of measurement must be ratio, or more generally, data must be continuous. This, too, is a very restrictive condition, for the majority of data collected in applied areas of statistics are not continuous, but categorical which almost always contain nonlinear relations between variables. The situation surrounding the normal distribution is almost paradoxical, for it suggests that for us to deal with nonlinear relations, for example, we must give up the assumption of the normal distribution and the level of measurement must typically be lower than ratio.

MUNDA handles both nominal and ordinal measurement and can capture both linear and nonlinear relations between variables with relative ease. This capability of any method is much desired in practice, where data are collected without any rigorous sampling scheme or at the level of lower than ratio measurement. In the current book, we accept the fact that data are typically categorical, obtained without any sampling scheme, and that the data often contain nonlinear relations between variables. Thus, the object of scaling needs to be redefined, namely, not only upgrade the level of derived measurement, but also capture whatever information is embedded in them.

To this end, it is convenient to classify our categorical data into two types, incidence data and dominance data, which roughly correspond to Stevens' nominal and ordinal data, respectively, since as we see shortly they require different objectives and operations. For more thorough coverage of measurement topics, see Hand (1996, 2004).

Table 4.1 *Example: Laxatives and Effects*

Laxative	No Effect	Some	Right	Too Much	Total
A	0	3	6	21	30
B	5	15	9	1	30
C	2	18	10	0	30
Total	7	36	25	22	90

4.2 Classification of Categorical Data

Nishisato (1993) classified categorical data into two categories, incidence data and dominance data. When we obtain such data as contingency tables and multiple-choice data, we have entities (data elements) which can be weighted to generate a composite variable. This class of data is one type and is called incidence data. In contrast, when we have such data as rank-order and paired comparison data, we see no entities or definite pieces of information that can be weighted, for in paired comparisons we know only whether or not Stimulus 1 is preferred to Stimulus 2, without any information as to how much more one is preferred to the other. This class of data constitute the other type, called dominance data. Nishisato's classification takes into consideration the differences in the corresponding scaling procedures.

4.2.1 Incidence Data

Let us look at three data types which belong to incidence data. The elements of incidence data are either the presence, coded 1, or absence, coded 0, of an attribute, response frequencies as we will see next.

(i) Contingency Tables

See a numerical example in Table 4.1. The data were collected by asking each of three groups of 30 people each to judge one of three types of laxatives A, B and C. The data are joint response frequencies of two sets of categories, laxatives and ratings.

(ii) Multiple-Choice Data

Multiple-choice data are an extension of the contingency table of two categorical variables to more than two categorical variables. When there are more than two categorical variables, the corresponding contingency table is multidimensional. However, in practice it is not convenient to

Table 4.2 *Gun-Control Bylaw*

Subject	20-30	31-40	41-50	Yes	No	A	B	C	D
1	0	1	0	1	0	1	0	0	0
2	0	0	1	1	0	0	0	1	0
3	1	0	0	0	1	0	1	0	0
4	1	0	0	1	0	0	0	0	1
.	.	.	.	.	.	.	.	.	.
.	.	.	.	.	.	.	.	.	.
65	0	0	1	1	0	1	0	0	0

deal with a multidimensional contingency table because of its difficulty in expressing it as a table and because of the possibility of many empty cells in the table. Instead, therefore, we represent the data in the form of respondents-by-categories of multiple-choice questions. See the example in Table 4.2, in which the following three multiple-choice questions were asked and three response-pattern matrices are arranged in a row.

Q.1: What is your age group? [20-30;31-40;41-50]
Q.2: Do you support the new gun control bylaw? [yes, no]
Q.3: Which districts do you live in? [A,B,C,D]

Notice that this "response-pattern table" consists of 1s and 0s to indicate choices and non-choices, respectively. It is typically the case that subjects choose only one category per question.

(iii) **Sorting Data**
There was a time when sorting data were popular among cognitive psychologists because of the freedom which subjects can exercise in judging (Nagy, 1984). This data type, however, is not one of the widely collected data types, and some explanation is in order. Suppose that the following study subjects were presented to a group of students:

A=English, B=History,C=Mathematics, D=Physics,
E=Psychology, F=Biology, G=Education.

Suppose students were asked to give 1 to the first study subject presented to them and give 1 to each of the remaining study subjects

Table 4.3 *Sorting of Study Subjects into Piles*

Student	1	2	3	4	5	6	7	8
A	1	1	2	3	4	3	1	2
B	1	2	2	3	3	3	1	1
C	2	3	1	2	2	2	2	3
D	2	3	1	2	2	2	2	3
E	3	4	2	1	1	3	1	4
F	4	4	2	2	5	1	2	5
G	1	1	2	1	1	3	1	1

which they think is similar to the first one. Once this task is over, give 2 to the first study subject from the remaining subjects, and go down the list to identify those which are similar to it and give them 2s. This task should be repeated until all the study subjects are sorted into similar piles. The decision on both the number of piles and the size of each pile is completely up to the students, the reason why sorting data are said to be comparatively free from the restrictions on judgment. Sample data are given in Table 4.3, where eight students sorted seven study subjects into piles in terms of similarity.

 In processing the collected data, the above table will be converted into the (study subjects)-by-piles table of response patterns, that is, 1s and 0s. Therefore, the sorting data table is very much like that of multiple-choice table, except the rows and the columns are interchanged: in sorting data, respondents are listed in columns with piles treated as if they were response categories, and objects to be sorted are arranged in rows. Thus, one respondent who sorted objects into 5 piles is treated here as if the person had 5 response categories. In our example, the responses for A in the first row [11234312] are converted to [(1000),(1000),(01),(001), (00010),(001),(1000), (01000)] because respondents 1, 2, 3, 4, 5, 6, 7 and 8 created 4, 4, 2, 3,5, 3, 4, and 5 piles, respectively.

Characteristics of Incidence Data
When data are subjected to MUNDA:

1. There exists a trivial solution (to be discussed shortly).
2. Chi-square metric is used to indicate distance relations.
3. Even when data are perfectly internally consistent (to be discussed shortly), data typically contain more than one component. In other

words, the data cannot generally be exhaustively explained by one component, except for a $2 \times n$ or an $m \times 2$ contingency table, or a larger table with redundant rows or columns.

4. Scaling is to find a low-rank approximation to the elements of the data matrix.

These are all technical matters that require further discussion. Let us therefore wait for awhile until we discuss details of MUNDA.

4.2.2 Dominance Data

The elements of data are ordinal measurements, and the objective of quantification is different from that of incidence data.

(i) **Paired Comparison Data**

Paired comparison data are obtained by asking subjects which one in the pair of objects they like better or consider more important. The response of Subject i to the paired comparison of two objects (X_j, X_k) is coded as

$$_i f_{jk} = \begin{cases} 1 & \text{if } X_j > X_k \\ 0 & \text{if } X_j = X_k \\ -1 & \text{if } X_j < X_k \end{cases} \tag{4.1}$$

Since the handling of -1 is not easy for data entry, we typically use the coding of 1, 2 and 0 for the preference for the first object, for the second object and the tied response, respectively, that is, we change -1 to 2 in the above coding. The coding of 1, -1 and 0, however, is the one used for analysis.

Suppose that five subjects were asked to judge which object in each of the following pairs they prefer: Pair A: (apple, cherry), B:(apple, mango), C:(apple, grapes), D:(cherry, mango), E:(cherry, grapes), F:(mango, grapes). Sample data are as in Table 4.4.

(ii) **Rank-Order Data**

Rank-order responses are coded as 1, 2, 3 and so on, where "1" indicates the first choice or the most preferred object, and the highest number the last choice. Let us look at a numerical example (Table 4.5), in which five managers A, B, C, D and E ranked seven applicants for a job. Tied ranks are given the average rank. For example, if the first two choices are tied, each gets 1.5; if the first three choices are tied, each gets 2. In this way, the sum of all ranks is fixed and equal to $\frac{n(n+1)}{2}$.

(iii) **Successive Categories Data**

Successive categories data are in essence the same as multiple-choice data, except for the following aspect: the same set of successively

Table 4.4 *Pair Comparisons of Four Fruits*

Pair	A	B	C	D	E	F
1	1	2	2	1	0	1
2	2	2	2	1	1	1
3	2	2	1	1	0	1
4	2	2	2	2	2	2
5	1	1	1	1	2	2

Table 4.5 *Ranking of Job Applicants*

Applicant	1	2	3	4	5	6	7
A	3	6	5	4	1	7	2
B	2	7	5	4	3	6	1
C	2	5	6	3	4	7	1
D	3	7	4	5	1	6	2
E	4	6	7	5	2	3	1

ordered categories is used for the judgments of all objects. For instance, consider using the following set of categories: 1 = Low, 2 = Medium, 3 = High, and ask five subjects to judge the prices of potato, banana, tomato and asparagus. Sample data are as in Table 4.6.

Why is this table not called multiple-choice data? There is no difference between successive categories data and multiple-choice data in terms of the content of information, but the difference lies in how to process the data. In the case of successive categories data, we will use a rather restricting mode of analysis in such a way that as in the case of the

Table 4.6 *Rating of Price of Vegetables*

Vegetable	Potato	banana	tomato	asparagus
1	1	1	3	3
2	1	1	2	3
3	2	1	3	3
4	2	2	2	2
5	1	1	2	2

traditional Thurstonian procedure of unidimensional analysis (Bock and Jones, 1968) we will determine not only the scale values of vegetables but also the so-called category boundaries, one between "low" and "medium" and one between "medium" and "high." With this added work, the above data are converted into rank-order data of category boundaries and objects. The detailed procedure will be discussed later in Chapter 12. At the current moment, we simply state that successive categories data are classified as dominance data because of the conversion into rank-order data.

Characteristics of Dominance Data Since we coded responses crisply rather than in inequality relations, we must opt for a compromised procedure, that is, we will treat ordinal numbers as cardinal. However, we will not compromise the objective of analysis that we determine the configuration of subjects and objects in such a way that it retains the original ordinal information.

Right now, let us look at some characteristics of analysis of dominance data.

1. There is no trivial solution.

2. Typically Euclidean metric is used, but note that there is nothing intrinsic about this property - if missing responses are involved in dominance data, thus making the marginals no longer constant, we must deal with the chi-square metric.

3. When data are perfectly internally consistent, only one component is sufficient to describe data, unlike the case of incidence data.

4. Scaling is to find a multidimensional configuration of row variables and column variables such that ordinal information in data is best approximated in the smallest dimensional space. This object is very different from that for incidence data.

The above distinction between the data types will become clear once we start looking at scaling of a variety of data types. Dual scaling was developed for these two types of data from its debut in 1969, while correspondence analysis and multiple correspondence analysis were first formulated only for incidence data, and not for dominance data. As mentioned in Chapter 3, however, researchers in correspondence analysis (e.g., Greenacre, Torres-Lacomba, van de Velden) have recently derived formulations of correspondence analysis for dominance data as well, thus diminishing the initial differences between dual scaling and correspondence analysis, except for some details.

4.3 Euclidean Space

This is a familiar medium in which we talk about "distance" in everyday life, in particular, three-dimensional Euclidean space. When we have two points x and y with coordinates in n-dimensional space, that is, $x = (x_1, x_2, ..., x_n)$ and $y = (y_1, y_2, ..., y_n)$, then the distance in the Euclidean space, or simply, the Euclidean distance $d(x, y)$ is defined as

$$d(x, y) = \sqrt{\sum_{i=1}^{n}(x_i - y_i)^2} \qquad (4.2)$$

In the Euclidean space, it is the common knowledge that the variance of a set of variables is non-negative because the norm of a vector cannot be smaller than the norm of its projection to any subspace. Notice that the variance is proportional to $\mathbf{x}'\mathbf{x} - \mathbf{x}'\mathbf{P}\mathbf{x}$, where $\mathbf{x}$ is a vector of variables and $\mathbf{P}$ is a projection operator. In non-Euclidean space, the norm of $\mathbf{P}\mathbf{x}$ can become larger than the norm of $\mathbf{x}$, resulting in a negative variance. When we consider two-dimensional space, we have the familiar theorem.

4.3.1 Pythagorean Theorem

The sum of the squares of the lengths of the legs of a right triangle is equal to the square of the hypotenuse. Geometrically, this means that the area A is equal to the sum of the areas B and C as shown in Figure 4.1. This theorem can be extended to any number of dimensions, and its general expression is the definition of the Euclidean distance in n-dimensional space, mentioned above. For two-dimensional space, we have another useful relation, the cosine law.

4.3.2 The Cosine Law

Consider three points i, j, k in two-dimensional Euclidean space. Let us indicate the distance between two points by d with two points in the subscript, and the angle between i, j and k with i at the origin by θ_{jik}. Then, there exists the relation,

$$d_{jk}^2 = d_{ij}^2 + d_{ik}^2 - 2d_{ij}d_{ik}\cos\theta_{jik} \qquad (4.3)$$

We will see that the product-moment correlation between two variables can be expressed as the cosine of the angle between two axes, each axis representing one variable, and covariance as the product of two standard

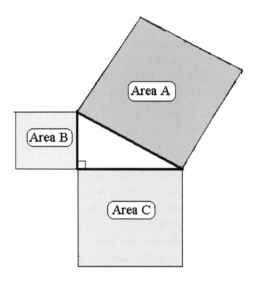

Figure 4.1 *Pythagoras theorem*

deviations times the correlation. Thus,

$$r_{jk} = \cos\theta_{jik} = \frac{d_{ij}^2 + d_{ik}^2 - d_{jk}^2}{2d_{ij}d_{ik}} \tag{4.4}$$

and covariance s_{jk} is given by

$$s_{jk} = d_{ij}d_{ik}\cos\theta_{jik} = \frac{d_{ij}^2 + d_{ik}^2 - d_{jk}^2}{2} \tag{4.5}$$

We will see shortly an application of the cosine law to the calculation of the so-called "between-set distance."

4.3.3 Young-Householder Theorem

This theorem was presented by Torgerson (1952).

1. If the matrix with the typical element being $d_{ij}d_{ik}\cos\theta_{jik}$, that is, the covariance matrix **V**, is positive definite (i.e., all eigenvalues are positive) or positive semi-definite (i.e., all eigenvalues are non-

negative, some may be zero), the distances between two points may be considered as distances between points in Euclidean space.

2. The rank of this matrix is equal to the dimensionality of space which accommodates all the points.

3. Any positive definite or positive semi-definite covariance matrix can be factored into the product of the matrix of coordinates of the points lying in Euclidean space. Namely,

$$\mathbf{V} = \mathbf{G}\mathbf{\Lambda}\mathbf{G}' = (\mathbf{G}\mathbf{\Lambda}^{\frac{1}{2}})(\mathbf{G}\mathbf{\Lambda}^{\frac{1}{2}})' = \mathbf{A}\mathbf{A}' \tag{4.6}$$

where

$$\mathbf{G}'\mathbf{G} = \mathbf{I}, \quad \mathbf{\Lambda} = diag(\lambda_i), \quad \lambda_i \geq 0 \tag{4.7}$$

Thus, it is useful to examine if all eigenvalues are non-negative to be sure that we are dealing with Euclidean space.

4.3.4 Chi-Square Distance

As noted earlier, categorical data can be grouped into two groups, and two different metrics are used to quantify them, the chi-square metric for incidence data and the Euclidean metric for dominance data. We should note, however, that both metrics are defined in Euclidean space.

Let us now look at the chi-square distance. In quantification of a two-way table, we determine row weights and column weights, and plot them as points in the space. Suppose that row j and row k have $f_{j.}$ and $f_{k.}$ observations each. We then indicate by $p_{j.}$ and $p_{k.}$ the respective proportions over the total responses f_t. Let us indicate by y_{js} and y_{ks} the scale values (i.e., weights) of the two respective rows in component s. Then, the squared chi-square distance between row points j and k is given by

$$d^2_{chi(jk)} = \sum_{s=1}^{K} \rho_s{}^2 \left(\frac{y_{js}}{\sqrt{p_{j.}}} - \frac{y_{ks}}{\sqrt{p_{k.}}} \right)^2 \tag{4.8}$$

Similarly, if we consider the scale values of column points, the squared chi-square distance between column points c and d is given by

$$d^2_{chi(cd)} = \sum_{s=1}^{K} \rho_s{}^2 \left(\frac{x_{cs}}{\sqrt{p_{.c}}} - \frac{x_{ds}}{\sqrt{p_{.d}}} \right)^2 \tag{4.9}$$

The above two expressions are what we call "within-set" distances. Recall that in our geometric approach to MUNDA in Chapter 2, we have already used the chi-square metric to calculate projected weights.

In analysis of two-way tables, our main interest lies not only in the within-set distances, but also in the between-set distances, that is, the distance between row i and column j. As we will see later, however, the space for row points and the space for column points are not the same, but are generally different. We will see later that our quantification method leads to the transformation of data in such a way that the discrepancy between row space and column space be a minimum. Because of our practice of plotting row points and column points as if they lie in the same space, we must caution people to be careful about interpreting a graph in which row points and column points are plotted together, for they span different spaces. To overcome this difficulty, Nishisato and Clavel (2002) presented the formula for the chi-square distance between row j and column c, that is, the between-set distances.

$$d^2_{chi(jc)} = \sum_{s=1}^{K} \rho_s{}^2 \left(\frac{y_{js}{}^2}{p_{j.}} + \frac{x_{cs}{}^2}{p_{.c}} - 2\rho_s \frac{y_{js}x_{cs}}{\sqrt{p_{j.}p_{.c}}} \right) \qquad (4.10)$$

The last formula is based on the cosine law: Consider two points A and B in two-dimensional space, with the distance between the origin (O) and A being a and the distance between the origin and B being b, and the angle between AOB being θ. Then the cosine law states that the square of the distance between A and B is given by

$$d_{AB}{}^2 = a^2 + b^2 - 2ab \cos \theta \qquad (4.11)$$

The above formula for the between-set distance is an application of the cosine law. Consider a 3×3 contingency table which typically requires only two dimensions since we remove the trivial component prior to analysis. But, once we take into consideration the discrepancy between row space and column space, such data indeed require three dimensions, that is, two proper components and one related to the discrepancy in space. See an example in Nishisato and Clavel (2002). With dominance data, all marginals are typically equal. In this case, the chi-square distance is reduced to the Euclidean distance. When dominance data have missing responses, however, we face the chi-square distance.

4.4 Multidimensional Space

4.4.1 Pierce's Concept

Pierce (1961) made an interesting remark about the distribution of information in multidimensional space. Consider a circle of radius 1 and a concentric circle of radius $\frac{1}{2}$ inside of it. The area of a circle is πr^2, thus the area of the outer circle is π and that of the inner circle is $\frac{1}{4}\pi$. If we consider a sphere, the volume is $\frac{4}{3}\pi r^3$. So, $\frac{1}{8}$ of the volume of a sphere lies within a sphere of one-half diameter. We can now generalize this to the point that the volume of a hypersphere of n dimensions is proportional to r^n. Let us now consider a sphere of radius r and a sphere of radius $0.99r$. For a 1000-dimensional sphere, a fraction of 0.00004 of the volume lies in a sphere of 0.99 of the radius. Pierce (1961, p.170) states that "The conclusion is inescapable that in the case of a hypersphere of a very high dimensionality, essentially all of the volume lies very near the surface!"

The above remark is educational. If all variables are standardized and if the data are purely two-dimensional, we will see that all the data points lie on the circle of radius 1. If some points lie inside and away from this circle, it indicates that the data require more than two dimensions.

4.4.2 Distance in Reduced Space

It is a wide-spread practice in multidimensional data analysis (e.g., factor analysis, principal component analysis, MUNDA) that the investigator examines only the first few components. In other words, the multidimensional structure of data is looked at, investigated and interpreted in reduced space. This practice is often justified with such a view that the first few major components account for an important portion of data, or a view that we cannot sensibly interpret all the components. These views may be partially right, but can we indeed justify the practice in this way? Please note that this practice can pose a serious problem for validity of interpretation.

From the equation for the Euclidean distance, it is obvious that if two points lie in multidimensional space the distance between them increases as we view it in higher-dimensional space than two- or three-dimensional space. Thus, if two points are very far when viewed in two-dimensional space, there is a guarantee that they are further apart from each other in three- or higher-dimensional space. However, the opposite is not true. If two points are very close to each other in two-dimensional space, there is no guarantee that they are also close in three- or higher-dimensional space.

This appears contrary to the general practice in multidimensional data analysis. People typically look for a cluster of points which are *close to one another* in two- or three-dimensional space. Why? There is no guarantee that those close points form a cluster in high-dimensional space. Thus, we should instead look for points in two- or three-dimensional space which are *widely separated* from one another, for they are widely separated in multidimensional space, too. This argument favors the procedure of defining clusters in multidimensional space, rather than in reduced space.

For many years, it has been the tradition especially among those users of factor analysis to interpret each "factor," that is, how variables cluster around each axis. Because of this, researchers did not question the practice of interpreting data in reduced space. This practice, however, should be questioned in MUNDA for the reason stated above. Each component is a principal axis, which is determined by all the variables in the data set. Thus, in a situation in which the researcher suspects multi-dimensionality of an attribute such as "anxiety," he or she may introduce as many items relevant to anxiety as possible. This attempt in response to the original inquiry about multidimensionality of anxiety has the effect of changing principal axes, which may have the effect of diluting otherwise clear-cut identities of distinct forms of the variety of anxiety. It is likely, however, that clustering of variables remain more or less the same with an increased number of variables in the data set, although such clusters may no longer fall on principal axes. This leads to the question of "axis interpretation" versus "cluster interpretation." The present argument appears to favor the cluster interpretation, made in multi-dimensional space, over the axis interpretation of dimension by dimension. We will discuss this problem later again.

4.4.3 Correlation in Reduced Space

Similar to the above problem or issue of multidimensional space is the interpretation of correlation between two variables. When two variables are expressed as two vectors with the origin in two-dimensional graph, it is well known that the Peasonian correlation is equal to the cosine of the angle between the two vectors. When we express data in the orthogonal coordinate system, it is known that each variable can be expressed as an axis, along which all variates of the variable lie. Suppose that two variables lie in three-dimensional space, and that we look at them in two-dimensional space. It is not difficult to see that the angle between two axes, viewed in two-dimensional space, cannot be larger than the angle between them, viewed in three-dimensional space (Nishisato, 2005c). In other words, when two variables lie in multidimensional space with angle θ, this angle becomes smaller when we project the two axes onto a space of smaller dimensionality than that, for example, two-dimensional space. This means that if we look at the data in two dimensions when they are actually in a space of dimensionality higher than two, the cosine of the smaller angle, viewed in two dimensions, is larger than the cosine of the original larger angle. Thus, looking at data in reduced space has the effect of magnifying correlation, that is, over-estimation of correlation between variables. Almost every researcher interprets data in reduced dimension, but shouldn't this practice create problems? Should we not be concerned that when all variables have negligible inter-variable correlation they still look highly correlated if viewed in two-dimensional space?

The above point may appear paradoxical when we consider the model of linear factor analysis, in which the correlation between two variables is decomposed into the contributions of many dimensions, giving us the impression that the more dimensions involved the higher the correlation between variables. This, however, is not the case as the previous paragraph clearly explains. The topic of interpreting data in reduced space is indeed a serious problem, yet it has been accepted as a well-established procedure for multidimensional data analysis. In the last chapter, we will discuss this topic as a tricky problem in multi-dimensional data analysis.

CHAPTER 5

Technical Preliminaries

In Chapter 4, we looked at concepts which may be of use. In this chapter, some technical information is presented primarily for those in applied areas of statistics.

5.1 Linear Combination and Principal Space

Suppose that scores on two tests, X_1 and X_2, are highly correlated. Then it would be convenient if we can combine the two scores and generate a single score (composite score). It is desirable that such a composite score contains as much information of the original two tests as possible. Let us express a composite score as a sum of weighted scores of two tests, that is, the composite score, y, as a (weighted) linear combination of the two tests:

$$y_i = w_1 x_{1i} + w_2 x_{2i} \tag{5.1}$$

for Subject i. It is a well accepted procedure to choose the weights in such a way that

$$w_1{}^2 + w_2{}^2 = 1 \tag{5.2}$$

The weights, satisfying (5.2), are called *normalized weights*. Suppose we consider a linear combination of *n* variables. The normalized weight for variable *i*, say w_i, can be obtained from initial weights, u_i, by the following formula:

$$w_i = \frac{u_i}{\sqrt{\sum_{j=1}^{n} u_j{}^2}} \tag{5.3}$$

This normalization of weights plays three important roles in data analysis:

1. Preserving the unit of measurement

2. Projecting data points on the chosen axis (space)

3. Choosing principal axes (principal space)

Unit Preservation: Consider scores of five subjects on Test 1 and those of the same five subjects on Test 2 to be respectively [6, 8, 4, 10, 2] and

[3, 4, 2, 5, 1]. It is obvious that $x_{1i} = 2 \times x_{2i}$, that is, every subject's score on X_1 is twice the score on X_2. Thus if we plot five subjects with two scores as coordinates for axes X_1 and X_2, all the subjects lie on the straight line, indicated by axis Y. The position of Subject 1, for example, can be calculated by the Pythagoras theorem as the square root of the sum of squares of 6 and 3, that is, the square root of (36+9). A formula to calculate the position of each subject in this example can be derived as follows:

1. Express the composite axis Y as one passing through the origin $(0, 0)$ and $(2, 1)$, and the composite score as an equation

$$y_i = u_1 x_{1i} + u_2 x_{2i} = 2x_{1i} + x_{2i} \tag{5.4}$$

2. Normalize the two weights 2 and 1 to w_1 and w_2, respectively, by the formula discussed earlier.

In the present example, the final expression is given by

$$y_i = \frac{2}{\sqrt{5}} x_{1i} + \frac{1}{\sqrt{5}} x_{2i} \tag{5.5}$$

Thus, the composite scores for the five subjects are, from Subject 1 to Subject 5, 6.71, 8.94, 4.47, 11.18, 2.24, respectively.

Projection of Any Point on the Chosen Axis: Suppose that we decide to use the above composite axis Y, and that Subject 6, who scored 7 and 6 on X_1 and X_2, respectively, was added to the original five subjects. Then, what composite score should this subject get? Or, rather, what is the projection of the point (7,6) on axis Y? In Figure 5.1, $A*$ is the data point with coordinates (7,6), and the axis onto which we would like to project the data point goes through the origin and the point (a,b), where a and b can be any points on the axis such as (2,1),(4,2),(8,4) and (10,5). We are interested in the length from the origin O to the projected point A, which is the composite score of the person whose scores on two tests are 7 and 6. Therefore, if we use (2,1) for (a,b),

$$\overline{OA} = \overline{OB} + \overline{BA} = \overline{OB} + \overline{B*A*} = \overline{OC}\cos\theta + \overline{A*C}\sin\theta$$
$$= \overline{OC}\frac{a}{\sqrt{a^2 + b^2}} + \overline{A*C}\frac{b}{\sqrt{a^2 + b^2}} = 7\frac{2}{\sqrt{5}} + 6\frac{1}{\sqrt{5}} = 8.9. \tag{5.6}$$

Thus, the composite score y_6 is 8.9. In this way, we can calculate projection of any points on the axis going through the origin and point (a,b). The important point here is that the weights must be normalized.

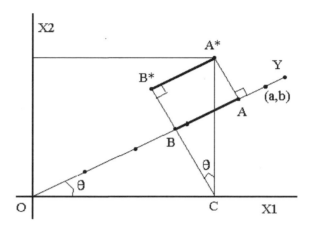

Figure 5.1 *Projection of Data Points on an Axis*

Choice of Principal Axes: Generally speaking, two data sets (X_1, X_2) would not reveal such a scatter plot as a straight line, but a scatter with a much greater variation. Thus the determination of the composite axis is not as straight a problem as seen above, but becomes a point of great importance. Since the normalization of weights allows us to obtain projections of data points on a specific axis, we can reverse the task to that of determining the composite axis so that the composite scores have a certain property. Since our interest lies in producing composite scores that are as representative of the original two sets of scores as possible, we can state the task as that of determining weights w_1 and w_2 so as to maximize the variance of the composite scores. In other words, we should determine w_1 and w_2, subject to the normalization condition, such that the variance of *y* be a maximum.

The composite axis with a maximal value of variance is called the principal axis, and the variance of the composite scores (projections of data points on the principal axis) is called an eigenvalue. Principal component analysis (PCA) is a mathematical procedure to derive composite scores as projections onto principal axes, and as such it is often referred to as eigenvalue decomposition of data.

The above case of two variables can easily be generalized to the case of n variables. Given n variables $X_1, X_2, ..., X_n$, define composite score y as a linear combination of the variables, that is,

$$y = w_1 X_1 + w_2 X_2 + \cdots + w_n X_n, \tag{5.7}$$

where $w_j, j = 1, 2, \cdots, n$, are normalized weights. The (first) principal component scores are $y_i, i = 1, 2, \cdots, N$ of N subjects that provide the maximal possible variance, and the weights w_j are called loadings. Supposing that N is much larger than n, PCA typically identifies n principal components. Thus, typical outputs of PCA of an $N \times n$ data matrix are n sets of N principal component scores, and n sets of n loadings for n variables. The first principal component is given by the composite scores of the maximal variance; the second principal component is associated with the composite scores of a maximal variance under the condition that it is independent of the first component (i.e., once the first component is extracted, remove the contribution of the first component from data, and subject the residual matrix to PCA to determine weights so as to maximize the variance of the composite scores); the third component maximizes the variance of the composite scores, on the condition that it is independent of the first two components, and so on.

Using the normalization condition as the constraint, a typical formulation of PCA can be stated as a solution to the maximization problem:

$$\frac{\partial L(w_j, \lambda)}{\partial w_j} = 0, \frac{\partial L(w_j, \lambda)}{\partial \lambda} = 0, \tag{5.8}$$

where $L(w_j, \lambda)$ is the Lagrangian function, given by

$$L(w_j, \lambda) = (\text{sum of squares of } y_i) - \frac{1}{2}(\sum_{j=1}^{n} w^2{}_j - 1) \tag{5.9}$$

When each variable is centered, that is, 0 origin, this maximization task leads to the eigenequation

$$(\mathbf{V} - \lambda\mathbf{I})\mathbf{w} = \mathbf{0} \tag{5.10}$$

or, if the variables are standardized,

$$(\mathbf{R} - \lambda\mathbf{I})\mathbf{w} = \mathbf{0} \tag{5.11}$$

where $\mathbf{V}$ and $\mathbf{R}$ are the variance-covariance matrix and the correlation matrix, respectively. The vector of loadings, say $\mathbf{a}$, is given by

$$\mathbf{a} = \frac{1}{\sqrt{\lambda}}\mathbf{Xw} \qquad (5.12)$$

The space specified with principal axes is called the principal space. We will hear about such expressions as the principal plane, principal hyper-plane, principal hyper-sphere and the principal hyper-space, all being defined in terms of principal axes as coordinate systems.

5.2 Eigenvalue and Singular Value Decompositions

As mentioned in the previous chapter on the history of quantification theory, *eigenvalues* emerged a long time ago as a topic of mathematical interest. Nowadays, eigenvalues (latent roots, characteristic roots) appear in many contexts. For instance, consider a quadratic function of x and y and we want to eliminate the product term of both variables. Suppose the quadratic function is given by

$$f(x,y) = 5x^2 + 8xy + 5y^2 = 9 \qquad (5.13)$$

The quadratic equation without the product term is called the canonical form, which is obtained by the procedure called canonical reduction. In the present example, the canonical form is given by

$$f(x,y) = 9x^2 + y^2 = 9 \qquad (5.14)$$

These two expressions of the same quadratic equation are related by orthogonal transformation as can be seen in the corresponding graphs with the original axes X, Y and the principal axes X^*, Y^*.

The two coefficients of the canonical form, 9 and 1, are called eigen-values. Notice that the quadratic function is completely symmetric with respect to the principal axes. In passing, it should be mentioned that the word 'dual' of dual scaling (Nishisato, 1980a) also means 'symmetric.'

The canonical reduction of a quadratic form is obtained by solving an eigenequation of the form,

$$\begin{bmatrix} 5 & 4 \\ 4 & 5 \end{bmatrix} \begin{bmatrix} x \\ y \end{bmatrix} = \lambda \begin{bmatrix} x \\ y \end{bmatrix}, \quad \text{or} \quad \mathbf{Aw} = \lambda\mathbf{w} \qquad (5.15)$$

where λ is the eigenvalue. The above eigenequation yields two eigen-values 9 and 1 and the corresponding eigenvectors $\mathbf{w}_1$ and $\mathbf{w}_2$. If we indicate the diagonal matrix of eigenvalues by $\mathbf{\Lambda}$ and the matrix of two eigenvectors $[\mathbf{w}_1, \mathbf{w}_2]$ by $\mathbf{W}$, the canonical reduction amounts to the transformation $\mathbf{W'AW} = \mathbf{\Lambda}$.

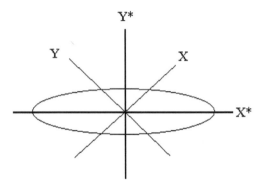

Figure 5.2 *Quadratic equation and canonical form*

Consider more generally a symmetric matrix $\mathbf{A}$. Then, the eigen-equation is given by

$$[\mathbf{A} - \lambda\mathbf{I}]\mathbf{w} = \mathbf{0}, \quad \text{or,} \quad \mathbf{A}\mathbf{w} = \lambda\mathbf{w} \tag{5.16}$$

where λ is an eigenvalue and $\mathbf{w}$ is the corresponding eigenvector. When $\mathbf{A}$ is an $n \times n$, there exists in general n eigenvalues and n corresponding eigenvectors. For this entire set, we have

$$\mathbf{A}\mathbf{W} = \mathbf{W}\boldsymbol{\Lambda} \tag{5.17}$$

where $\mathbf{W} = [\mathbf{w}_1, \mathbf{w}_2, \cdots, \mathbf{w}_n]$, and $\boldsymbol{\Lambda} = diag[\lambda_1, \lambda_2, \cdots, \lambda_n]$. Since $\mathbf{W}$ is an orthogonal matrix, its inverse is equal to the transpose. Therefore, premultiplying the complete eigenequation by $\mathbf{W}'$, we obtain

$$\mathbf{W'AW} = \mathbf{\Lambda} \qquad (5.18)$$

Thus, the eigenequation can be viewed as *orthogonalization* of the original matrix $\mathbf{A}$, or $\mathbf{A}=\mathbf{W\Lambda W'}$. Because $\mathbf{W}$ is an orthogonal matrix, it follows that

$$\mathbf{A}^k = \mathbf{AA}...\mathbf{A} = (\mathbf{W\Lambda W'})(\mathbf{W\Lambda W'})...(\mathbf{W\Lambda W'}) = \mathbf{W\Lambda}^k\mathbf{W'}$$
$$(5.19)$$

since $\mathbf{W'W} = \mathbf{I}$, that is,

$$\mathbf{A}^k\mathbf{W} = \mathbf{W\Lambda}^k \qquad (5.20)$$

In other words, raising the power of $\mathbf{A}$ does not change the matrix of eigenvectors, but raises the power of each eigenvalue to k. This aspect is fully used in the power method such as MRA.

Singular value decomposition (SVD) is more general than eigenvalue decomposition (EVD) in the sense that SVD deals with any square or rectangular matrix, say $\mathbf{F}$. SVD is also based on the orthogonalization of a matrix, that is,

$$\mathbf{YFX'} = \mathbf{\Phi} \quad \text{such that} \quad \mathbf{Y'Y} = \mathbf{I}$$
$$\mathbf{X'X} = \mathbf{I}, \quad \text{and} \quad \mathbf{\Phi} = diag[\rho_1, \rho_2, \cdots, \rho_r] \qquad (5.21)$$

In other words, any matrix can be expressed as the product of an orthogonal matrix to depict the row structure, times a diagonal matrix of singular values, times an orthogonal matrix to describe the column structure of the data matrix,

$$\mathbf{F} = \mathbf{Y'\Phi X} \qquad (5.22)$$

The relation between SVD and EVD is simple. Consider the following two forms of the product of a rectangular matrix: (1)$\mathbf{FF'}$ and (2)$\mathbf{F'F}$

$$\mathbf{FF'} = (\mathbf{Y'\Phi X})(\mathbf{X'\Phi Y}) = \mathbf{Y'\Phi}^2\mathbf{Y} = \mathbf{Y'\Lambda Y}$$
$$\mathbf{F'F} = (\mathbf{X'\Phi Y})(\mathbf{Y'\Phi X}) = \mathbf{X'\Phi}^2\mathbf{X} = \mathbf{X'\Lambda X} \qquad (5.23)$$

So we obtain the well-known result that the square of a singular value is an eigenvalue and that the eigenvalues of $\mathbf{FF'}$ are the same as those of $\mathbf{F'F}$.

Both EVD and SVD are based on the idea of principal hyperspace, a rectangular coordinate system defined by principal axes. As we saw earlier, "principal" means "symmetric," and, as many researchers have demonstrated so far, the decomposition of data in terms of principal axes possesses the least-squares property (for example, see Keller (1962), Johnson (1963), Schönemann, Bock and Tucker (1965)).

5.3 Finding the Largest Eigenvalue

In dealing with SVD or EVD, we often use an iterative method to extract singular values or eigenvalues according to the order of their magnitudes. This is so because in most multidimensional analysis we are not interested in all the components but only in a few components with large singular values or eigenvalues. To this end, let us discuss how to extract the largest eigenvalue.

5.3.1 Some Basics

Let $\mathbf{V}$ be an $n \times n$ symmetric matrix (i.e., $\mathbf{V}=\mathbf{V}'$), $\mathbf{\Lambda}$ be the $n \times n$ diagonal matrix of the eigenvalues (latent roots, characteristic roots, proper values), and $\mathbf{U}$ be the $n \times n$ matrix with the eigenvectors (latent vectors, characteristic vectors, proper vectors) in its columns. Then, the eigenequation is defined by

$$\mathbf{V}\mathbf{u}_j = \lambda_j \mathbf{u}_j \qquad \text{or} \qquad \mathbf{V}\mathbf{U} = \mathbf{U}\mathbf{\Lambda} \qquad (5.24)$$

Since $\mathbf{U}$ is an orthogonal matrix, its inverse is given by the transpose of $\mathbf{U}$, and it follows that

$$\mathbf{V} = \mathbf{U}\mathbf{\Lambda}\mathbf{U}' \text{ and } \mathbf{\Lambda} = \mathbf{U}'\mathbf{V}\mathbf{U} \qquad (5.25)$$

Suppose k is a positive integer. As we have already seen

$$\mathbf{V}^k = \mathbf{V} \times \mathbf{V} \times \mathbf{V} \times \cdots \times \mathbf{V}$$
$$= (\mathbf{U}\mathbf{\Lambda}\mathbf{U}') \times (\mathbf{U}\mathbf{\Lambda}\mathbf{U}') \times \cdots (\mathbf{U}\mathbf{\Lambda}\mathbf{U}') = \mathbf{U}\mathbf{\Lambda}^k\mathbf{U}' \qquad (5.26)$$

Consider any arbitrary vector of n elements, $\mathbf{b}_0$. This can be expressed as a linear combination of n eigenvectors, $\mathbf{u}_i$, that is,

$$\mathbf{b}_0 = \sum_{i=1}^{n} c_i \mathbf{u}_i. \tag{5.27}$$

But, since $\mathbf{V}\mathbf{u} = \lambda \mathbf{u}$, we obtain

$$\mathbf{V}\mathbf{b}_0 = \sum_{i=1}^{n} \lambda_i c_i \mathbf{u}_i. \tag{5.28}$$

Let us form a sequence of vectors as follows:

$$\mathbf{b}_0, \mathbf{V}\mathbf{b}_0 = \mathbf{b}_1, \mathbf{V}\mathbf{b}_1 = \mathbf{V}^2\mathbf{b}_0 = \mathbf{b}_2, \cdots\cdots,$$
$$\mathbf{V}\mathbf{b}_{p-1} = \mathbf{V}^p\mathbf{b}_0 = \mathbf{b}_p \tag{5.29}$$

$$\mathbf{b}_p = \mathbf{V}^p\mathbf{b}_0 = \sum_{i=1}^{n} \lambda_i{}^p c_i \mathbf{u}_i$$

$$= \lambda_1{}^p \left\{ c_1 \mathbf{u}_1 + \sum_{j=2}^{n} \left\{ \frac{\lambda_j}{\lambda_1} \right\}^p c_j \mathbf{u}_j \right\} \tag{5.30}$$

Therefore, assuming that eigenvalues are distinct and arranged from the largest λ_1 to the smallest λ_n, $\frac{\lambda_j}{\lambda_1}$, (j = 2,3,..., n), is a fraction, hence

$$\lim_{p \to \infty} \mathbf{b}_p = \lambda_1{}^p c_1 \mathbf{u}_1, \quad , \text{and} \quad \lambda_1 = \frac{\mathbf{b}_p' \mathbf{b}_p}{\mathbf{b}_p' \mathbf{b}_{p-1}} \tag{5.31}$$

To extract the second component, we calculate the residual matrix $\mathbf{V}_1$ by eliminating the contribution of the first component from $\mathbf{V}$,

$$\mathbf{V}_1 = \mathbf{V} - \lambda_1 \frac{\mathbf{u}_1 \mathbf{u}_1'}{\mathbf{u}_1' \mathbf{u}_1} \tag{5.32}$$

The maximal eigenvalue of $\mathbf{V}_1$, λ_2, and the associated eigenvector $\mathbf{u}_2$ are then calculated from $\mathbf{V}_1$.

5.3.2 MRA Revisited

Let $\mathbf{V}$ be an $n \times n$ symmetric matrix, and $\mathbf{b}_0$ be an $n \times 1$ arbitrary non-null vector. k_j is the largest absolute value of a resultant vector $\mathbf{b}_j$. Form the following sequence which is known to be mathematically convergent

(Nishisato, 1980a):

$$\mathbf{V}\mathbf{b}_0 = \mathbf{b}_1, \quad \frac{\mathbf{b_1}}{k_1} = \mathbf{a}_1, \quad \mathbf{V}\mathbf{a}_1 = \mathbf{b}_2, \quad \frac{\mathbf{b_2}}{k_2} = \mathbf{a}_2, \cdots,$$

$$\mathbf{V}\mathbf{a}_{j-1} = \mathbf{b}_j, \quad \frac{\mathbf{b_j}}{k_j} = \mathbf{a}_j \qquad (5.33)$$

This sequence eventually reaches the state that $\mathbf{a}_{j-1} = \mathbf{a}_j$. Then,

$$\frac{\mathbf{b}_j}{k_j} = \mathbf{a}_j \quad \text{means} \quad \mathbf{b}_j = k_j \mathbf{a}_j \qquad (5.34)$$

Then, it follows that

$$\mathbf{V}\mathbf{a}_{j-1} = \mathbf{V}\mathbf{a}_j = \mathbf{b}_j = k_j \mathbf{a}_j, \quad \text{and} \quad \mathbf{V}\mathbf{a}_j = k_j \mathbf{a}_j \qquad (5.35)$$

The above formulas and the convergent sequence, put together, indicate that k is the largest eigenvalue and $\mathbf{a}$ the corresponding eigenvector.
When $\mathbf{V}$ is not square or symmetric, Nishisato (1988c) has shown that the process converges to two generally distinct constants, say k_j and k^*_j. Then, the eigenvalue is the product of these and the singular value is the geometric mean of the two constants, as seen in the numerical example of MRA.

5.4 Dual Relations and Rectangular Coordinates

Let $\mathbf{F}$ be a rectangular matrix of frequencies or $(1,0)$ incidences. Let us derive an optimal weight vector for its rows, $\mathbf{y}$, and one for its columns, $\mathbf{x}$, optimal in the sense that these vectors maximize $\mathbf{y}'\mathbf{F}\mathbf{x}$, subject to the constraint that $\mathbf{y}'\mathbf{D_r}\mathbf{y} = \mathbf{x}'\mathbf{D_c}\mathbf{x} = f_t$, where $\mathbf{D_r}$ is the diagonal matrix of row marginals of $\mathbf{F}$, $\mathbf{D_c}$ the diagonal matrix of column marginals of $\mathbf{F}$ and f_t the sum of elements of $\mathbf{F}$. This is nothing but the task of MUNDA, and it leads to an eigenequation. It is well known that optimal weight vectors thus derived can be expressed in the form of dual relations, or substitution formulas in French correspondence analysis,

$$\mathbf{y}_k = \frac{1}{\rho_k}\mathbf{D}_r^{-1}\mathbf{F}\mathbf{x} \qquad (5.36)$$

$$\mathbf{x}_k = \frac{1}{\rho_k}\mathbf{D}_c^{-1}\mathbf{F}'\mathbf{y} \qquad (5.37)$$

where ρ_k is the singular value associated with component k. This set of equations provide an expression for normed weight vectors or standard coordinates. Let us rewrite the above set of equations as follows:

$$\rho_k \mathbf{y}_k = \mathbf{D}_r^{-1} \mathbf{F} \mathbf{x} \tag{5.38}$$

$$\rho_k \mathbf{x}_k = \mathbf{D}_c^{-1} \mathbf{F}' \mathbf{y} \tag{5.39}$$

This set provides projected weight vectors or principal coordinates. Singular value ρ_k serves as a projection operator for dual relations, and $\rho_k \mathbf{y}_k$ is the projection of vector $\mathbf{y}$ onto the space of $\mathbf{x}$ and $\rho_k \mathbf{x}_k$ the projection of $\mathbf{x}$ onto the space of $\mathbf{y}$. In this context, we can understand Hirschefeld's simultaneous linear regressions, where $\rho_k \mathbf{y}_k$ and $\rho_k \mathbf{x}_k$ are regressions of $\mathbf{y_k}$ on $\mathbf{x_k}$ and that of $\mathbf{x_k}$ on $\mathbf{y_k}$.

It is important to note that the difference between normed and projected weights is not just a matter of norming. It is the projected weight that contains information about data, or more directly it is the singular value that determines the structure of multidimensional data. We can reconstruct the data matrix $\mathbf{F}$ from projected weights. Normed weights alone are not informative enough to reconstruct the data matrix from. Thus, it is absurd to plot only normed weights to interpret the outcome of quantification analysis.

This comparison between normed weights and projected weights highlights the key role that the singular value plays in multidimensional analysis. To emphasize the point, let us consider a set of standardized continuous variables, the correlation matrix of which is of rank 2. In this case, we can describe data in two-dimensional space in terms of two principal axes. Figure 5.3 shows a typical distribution of such data in which the first eigenvalue is considerably larger than the second one. Since all the variables are standardized, they are located at a distance of 1 from the origin, and the data points are distributed on the circle of radius 1. Their coordinates, however, are not normed weights (standard coordinates) but projected weights (principal coordinates). There seems to be some misconception that normed weights provide coordinates of points along this circle. It is wrong.

5.5 Discrepancy Between Row Space and Column Space

Let us recall the dual relations between row weighting and column weighting. When the singular value ρ is 1, we can obtain $\mathbf{y}$ from $\mathbf{x}$ and $\mathbf{x}$ from $\mathbf{y}$ by reciprocal means. This means that $\mathbf{y}$ and $\mathbf{x}$ span the same space. But, we have seen that the case of ρ equal to 1 is the trivial solution, and any associated results are typically discarded. Thus, in general we can

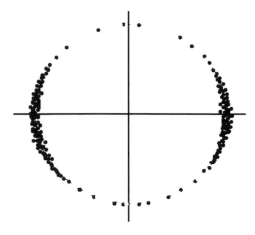

Figure 5.3 *Two-Dimensional Data and Projected Weights as Coordinates*

assume that ρ is not equal to 1. This is the situation in which the follow-ing warning applies to. "A great deal of caution is needed in interpreting the distance between a variable point and an individual point because these two points do not belong to the same space" (Lebart, Morineau and Warwick, 1984, p.19). Since our major interest lies in the joint display of rows and columns, this problem of discrepancy between row space and column space becomes a serious obstacle to our interest.

5.5.1 Geometrically Correct Joint Plots (Traditional)

The problem here is that we cannot justify a joint plot of projected row weights and projected column weights because they do not span the same space. Geometrically, we can plot normed row weights and projected column weights together, or projected row weights and normed column weights together. This choice eliminates the problem of discrepant row space and column space, and therefore it is the exact geometric representation. From the practical point of view, however, it is often the case that when the singular value is much smaller than its maximum of 1, the projected points tend to cluster close to the origin, thus making it difficult to compare the closeness of row points and column points, typically the main object of analysis to interpret data structure. This stems from a large difference between the norm of normed weights and that of projected weights.

5.5.2 Symmetric Scaling

To overcome this norm problem, we have the so-called "symmetric scaling," an option imported from France: Plot projected rows and projected columns in the joint display. This is the circumstance for which the warning by Lebart and others was made. Justification of this procedure seems to lie in the practice that most researchers examine only the first few components of which singular values are relatively large. As we will see later, however, even the largest singular value may be much smaller than 1, depending on data and data types. Nevertheless, symmetric scaling often provide easily interpretable graphs in practice, and we will see a number of numerical examples to attest this statement. It should be kept in mind that symmetric scaling is only a practical compromise.

5.5.3 CGS Scaling

Carroll, Green and Schaffer (1986, 1987, 1989) proposed the so-called CGS scaling, which stands for Carroll-Green-Schaffer scaling. Given a contingency table, they recommended to rearrange the data into the form of respondents (rows) by two sets of response categories, that is, into the response-pattern format of two multiple-choice data. Recall that if we express the responses to two questions in the form of (1, 0) response patterns, that is, F_1 (subjects-by-options of item 1) and F_2, then the contingency table is given by $F_1'F_2$. The CGS scaling is based on the analysis of $[F_1, F_2]$ so that all the options (categories) of two items are now represented by columns, hence they concluded that the between-set distance (option j of item 1 and option k of item 2) is now changed to the within-set distance since they are now both in columns. This met criticisms from Greenacre (1989). However, without resorting to Greenacre's criticism, one can look at a detailed comparison in data structure between the contingency table and the two-item response-pattern table (Nishisato, 1980a): whether one quantifies the response-pattern table or the corresponding contingency table, the weights for the columns of the response-pattern table are identical to the weights of rows and columns of the contingency table. Thus, we can conclude that the CGS scaling did not solve the problem of the row-column discrepancy. The CGS scaling works only when the singular value is 1. Otherwise, it is still the case in which the row space is overlaid to the column space, or vice versa.

5.5.4 Geometrically Correct Joint Plots (New)

AS mentioned earlier, Nishisato and Clavel (2002) discussed how to calculate the between-set distance. Their study is based on the point that the discrepancy between the row vector and the column vector of each component can be described by the two vectors being separated with an angle equal to the arc-cosine of the singular value, that is, $\cos^{-1} \rho$. Using this, they derived a formula for calculating the between-set distance, that is, the distance between a row and a column. They demonstrated that to accommodate data associated with the first two components one needs three-dimensional space, for the discrepancy between the row space and the column space contributes to the total space as an additional dimension. This means the following problems: if we use the geometrically correct joint plot (e.g., normed row weights and projected column weights), we lose the comparability of the two norms, meaning that the graph is distorted; if we use a symmetric joint plot, the space discrepancies are ignored. Either way is not satisfactory. The possibility of calculating the between-set distance opens up a new problem for MUNDA: we must develop a procedure to represent exact between-set distances and exact within-set distances in joint display. This is a very tricky problem from the practical point of view, but it is important enough to investigate for MUNDA procedure to become truly optimal. Nishisato and Clavel (2002) strongly recommend that we develop a procedure to analyze not only within-set distance relations but also between-set distance relations, that is, analysis of the matrix **D** where

$$\mathbf{D} = \begin{bmatrix} \mathbf{D}_{rr} & \mathbf{D}_{rc} \\ \mathbf{D}_{cr} & \mathbf{D}_{cc} \end{bmatrix} \tag{5.40}$$

where $\mathbf{D}_{rr}$ and $\mathbf{D}_{cc}$ are within-set distances and $\mathbf{D}_{rc}$ and $\mathbf{D}_{cr}$ are between-set distances. One may object to the analysis of this super matrix because it would not help our aim of dimension reduction. However, so long as our emphasis of data analysis is placed on exhaustive analysis of information, this should be the direction in which the future of MUNDA should advance. Our main interest in a two-way table is in the row-column relation. Thus, why should we leave out this important piece of information from analysis?

5.6 Information of Different Data Types

Although we must define information in data rigorously, we will adopt here a popular measure of information, that is, the sum of eigenvalues associated with our quantification task. This information depends on the data types, discussed in Chapter 4.

Since the six data types have already been illustrated with numerical examples, we will here simply specify the dimensions of those data matrices. Since sorting data are essentially the same as multiple-choice data, except for interchanged rows and columns, we will omit sorting data from the current discussion.

1. Contingency table: n rows and m columns
2. Multiple-choice data: N subjects, n items, with m_j options, $j = 1,2,...,n$, and $\sum m_j = m$
3. Rank-order data: N subjects, n ranked objects
4. Paired comparison data: N subjects, $\frac{n(n-1)}{2}$ pairs
5. Successive categories data: N subjects, n objects, $m+1$ categories

Nishisato (1993) has shown that the sums of eigenvalues, associated with quantification of these data, are different, as shown in Table 5.1. Different data types contain different amounts of information. Does this have any implications for interpreting quantification outcomes? Does it prevent us from analyzing mixed data types? These are very important questions, and we will discuss them in the last chapter. Study Table 5.1 as it is revealing the nature of data as seen from the quantification view.

We have discussed some conceptual and technical preliminaries. Let us now move on to look at quantification of different types of data. Although it may be redundant, each application chapter will start with a numerical example so that we may not have to go back to refresh our memory. Since each data type requires special considerations for quantification, some basics will be provided for each data type. As for fundamentals of MUNDA, refer to Chapter 2.

Table 5.1 *Data types and information*

Data Type	Sum of Eigenvalues
Contingency Table	$\frac{\chi^2}{f_t}$
Multiple-Choice Data	$\sum_{j=1}^{n} \frac{m_j}{n} - 1$
Rank-Ordered Data	$\frac{n+1}{3(n-1)}$
Paired Comparison	$\frac{n+1}{3(n-1)}$
Successive categories data	$\frac{n+m+1}{3(n+m-1)}$

PART II

Analysis of Incidence Data

Incidence data are characterized by presence (typically coded 1) or absence of a response (typically coded 0), or a frequency of a response. These quantities are the objects of quantification. In other words, each response or a frequency of a response is weighted in such a way that a certain statistic is optimized. Because frequencies are involved in quantification, the distance between two data points is what we call the chi-square metric - if all frequencies are equal, the chi-square metric reduces to Euclidean metric. The quantification of incidence data always involves a trivial solution (component) which optimizes the quantification, irrespective of the data and thus is discarded from analysis. In the current book, the following types of incidence data are discussed:

(1) Contingency Tables
(2) Multiple-Choice Data
(3) Sorting Data.

Recall the characteristics of incidence data as summarized in 4.2.1.

CHAPTER 6

Contingency Tables

6.1 Example

Contingency data are represented as a table of m rows by n columns of response frequencies or presence or absence of responses. The following is an example of a 3x6 table, which lists sales of three kinds of barley (A-C) at six super-markets in the past one month, the unit being a 500 gram bag.

Although the above example is a typical one, we also have the following

Table 6.1 *Contingency Table*

Barley	Store1	Store2	Store3	Store4	Store5	Store6
A	46	57	83	87	6	62
B	4	34	35	19	4	34
C	13	9	15	58	0	1

type. Suppose children were asked to check which features the animals they saw today at the zoo had. The quantification task is to determine weights for rows and those for columns of the table in such a way that joint frequencies are best explained.

Table 6.2 *Checking Features of Animals*

Feature	Large	Feather	Swim	Fly	Ferocious	Cute
Animal A	13	0	3	5	0	6
Animal B	1	30	3	30	8	2
Animal C	1	0	8	0	12	10
Animal D	16	0	4	0	2	18

6.2 Early Work

Hirschfeld (1935) posed the question if we can derive spacing of rows and spacing of columns of a two-way table in such a way that the regression of rows on columns and the regression of columns on rows be simultaneously linear. This led to the formulation, called the simultaneous linear regression approach (Lingoes, 1964). As mentioned earlier, however, Lingoes used this name to refer to Guttman's quantification approach and without any reference to Hirschfeld's study. Lingoes's nomenclature is quite accurate to describe Hirschfeld's method. Note that Hirschfeld's formulation yields the "dual relations," discussed earlier.

In 1941, Maung presented "Fisher's scoring method" for the contingency table, and provided four distinct formulations, which he demonstrated are mathematically equivalent. Some people give more credit to Fisher (1940) than to Maung (1941a) for the reason that Maung was working with Fisher and calling his method Fisher's. However, as far as the detailed formulations and presentations are concerned, the credit seems due more appropriately to Maung than to Fisher.

One of the four formulations is exactly the same as what Kendall and Stuart (1961) described as the canonical correlation approach to the quantification of contingency tables. It is interesting to note that in the study of structural equation modeling (abbreviated as SEM) researchers refer to one of the options for handling ordinal categorical measurement as the Kendall and Stuart canonical correlation approach. Historically, it should rather be the Maung canonical correlation approach.

Let us look at this canonical correlation approach. The contingency table is constructed from two categorical variables. Therefore, the data can be expressed in two different ways, the response-pattern format, say $[\mathbf{F}_1, \mathbf{F}_2]$, and the contingency table format, $\mathbf{F}_1'\mathbf{F}_2$, where $\mathbf{F}_j$ is the subjects-by-categories table of 1's and 0's of variable j. For instance, in Table 6.3, $\mathbf{F}_1$ is 9x3 and $\mathbf{F}_2$ is 9x4, and in Table 6.4, $\mathbf{F}_1'\mathbf{F}_2$ is 3x4. The canonical correlation is defined as the product-moment correlation between two linear composites, $\mathbf{F}_1\mathbf{y}$ and $\mathbf{F}_2\mathbf{x}$. This objective function and its maximization in terms of $\mathbf{y}$ and $\mathbf{x}$ leads to the generalized eigen-equation, which has already been discussed.

As Maung (1941a) demonstrated, both formats lead to identical optimal weight vectors. However, de Leeuw (1973) and Nishisato (1980a) showed that the eigen values are different, and they showed that the eigenvalue from the response-pattern matrix ρ_r^2 and the one from the

Table 6.3 *Response-Patterns Format*

Item	1			2				
Option	1	2	3	1	2	3	4	Sum
Sub 1	0	1	0	0	0	1	0	2
Sub 2	1	0	0	0	1	0	0	2
Sub 3	0	0	1	0	0	0	1	2
Sub 4	0	1	0	1	0	0	0	2
Sub 5	1	0	0	0	0	0	1	2
Sub 6	0	0	1	0	0	1	0	2
Sub 7	1	0	0	1	0	0	0	2
Sub 8	0	1	0	0	0	1	0	2
Sub 9	1	0	0	0	1	0	0	2
Sum	4	3	2	2	2	3	2	18

Table 6.4 *Contingency-Table Format*

	1	2	3	4	Sum
1	1	2	0	1	4
2	1	0	2	0	3
3	0	0	1	1	2
Sum	2	2	3	2	9

contingency table ρ_c^2 are related by

$$\rho_c^2 = (2\rho_r^2 - 1)^2 \tag{6.1}$$

Because of this exact relation between ρ_r^2 and ρ_c^2, one may still say that the formulations of the two data formats are mathematically identical. However, Nishisato (1980a) pointed out another difference: the response-pattern format generally yields more components than the contingency format. For instance, the response-pattern format in Table 6.3 yields five components, that is, the total number of categories (options) minus the number of variables (i.e., 7-2=5), while the contingency table in Table 6.4 provides two components, that is, the smaller of m and n minus 1 (i.e., 3-1=2). Our question then is "What are those extra components?" This is just a note to show that "mathematical equivalence" by Maung (1941a) thus does not mean everything being the same.

6.3 Some Basics

6.3.1 Number of Components

Consider a contingency table with m rows and n columns. The table always contains a trivial solution with the eigenvalue 1 and the weight vector $\mathbf{y}$ being $\mathbf{1}$ and $\mathbf{x}$ being $\mathbf{1}$, no matter what the table looks like, that is, independently of the data. This trivial solution corresponds to the portion of the data which can be explained by the frequencies expected when the rows and the columns are statistically independent, that is,

$$f_{ij} = \frac{f_{i.} f_{.j}}{f_t} \tag{6.2}$$

where

$$f_{i.} = \sum_{j=1}^{n} f_{ij}, \; f_{.j} = \sum_{i=1}^{m} f_{ij}, \; f_t = \sum_{i=1}^{m}\sum_{j=1}^{n} f_{ij} \tag{6.3}$$

Thus, the residual after removing this trivial solution is subjected to quantification. The maximal number of components, T(comp), is given by

$$T(comp) = min(n, m) - 1, \tag{6.4}$$

where $min(n, m)$ means the smaller value out of n and m. The decomposition of the contingency table can therefore be expressed as

$$\mathbf{F} - \frac{\mathbf{f}_r \mathbf{f}_c'}{f_t} = \frac{\mathbf{f}_r \mathbf{f}_c'}{f_t} [\rho_1 \mathbf{y}_1 \mathbf{x}_1' + \rho_2 \mathbf{y}_2 \mathbf{x}_2' + \cdots + \rho_K \mathbf{y}_K \mathbf{x}_K'] \tag{6.5}$$

where $K=min(m,n)-1$.

6.3.2 Total Information

The information in the contingency table is distributed over K non-trivial components. Following the popular practice, let us define the total information, $T(inf)$, as the sum of eigenvalues (or squared singular values or the sum of correlation ratios), excluding the contribution of the trivial component which is 1.

$$T(inf) = \sum_{k=1}^{K} \rho_k{}^2 = trace(\mathbf{D}_c{}^{-\frac{1}{2}} \mathbf{F}' \mathbf{D}_r{}^{-1} \mathbf{F} \mathbf{D}_c{}^{-\frac{1}{2}}) - 1 = \frac{\chi^2}{f_t} - 1 \tag{6.6}$$

As we know, the value of χ^2 depends on the data set. In other words, the amount of total information depends on a particular data set, and it can happen that it is not worth decomposing the data into many components. Whether or not to continue the analysis may be decided by looking at the χ^2 for the total table with degrees of freedom $(m-1)(n-1)$.

6.3.3 Information Accounted For By One Component

Now that we know how much information is distributed over $min(m,n)$-1 components, it is useful to know what proportion each component accounts for. We define the statistic δ percent as the percentage of the total information accounted for by a particular component,

$$\delta_j = \frac{100\rho_j{}^2}{\sum_{k=1}^{K}\rho_k{}^2} = \frac{100\rho_j{}^2}{T(inf)} \tag{6.7}$$

where K is the total number of components.

The above definition is popular in multivariate analysis. When it is applied to quantification of a contingency table, however, Nishisato (1996) cautioned the use of δ since it behaves in a strange way as a function of $T(inf)$. The problem is due to a peculiar nature of categorical data. To recapitulate his point, we should note that when the contingency table is square, say $m \times m$ and all the responses fall in the main diagonal positions, all K eigenvalues become identical and equal to one (the trivial component also has the eigenvalue of 1). This is a case of perfect coincidences among categories. The strange aspect lies in the fact that even then we need m-1 components to reproduce the data. This is strictly due to the fact that the rank of such a contingency table is m, one component being trivial. The statistic δ of each component in this case is equal to $\frac{100}{m-1}$ percent, rather than 100 percent.

In principal component analysis of continuous data, the above situation of perfect association corresponds to the case in which all values of inter-variable correlation are 1s, in which case only one component suffices to explain the data exhaustively. The above case peculiar to the contingency table becomes even bizarre, as pointed out by Nishisato (1996), that as the total information in the data becomes smaller (i.e., responses tend to distribute evenly over all cells) it is more likely to see that the distribution of eigenvalues shows steep declines from Component 1 to Component 2 and so on. This kind of decline of δ values from Component 1 to 2, 3 and so on can be observed in a totally different situation with continuous data, that is, when all the variables are

Table 6.5 *Biting Behavior of Pets*

	Not a biter	Mild Biter	Flagrant Biter	Row sum
Mice	20	16	24	60
Gerbils	30	10	10	50
Hamsters	50	30	10	90
Guinea Pigs	19	11	50	80
Column sum	119	67	94	280

highly correlated. Nishisato (1993) presented alternative statistics for δ.

$$\theta_k = \frac{100\rho_k^2}{min(n, m) - 1} \tag{6.8}$$

$$\gamma_k = \frac{100\rho_k^2}{1 + \sum \rho_i^2} \tag{6.9}$$

$$\nu_k = \frac{100\rho_k^2}{max(\rho_i^2)} = 100\rho_k^2 \tag{6.10}$$

where three denominators are respectively Cramér's upper bound for the chi-square divided by f_t, that is, $min(m, n) - 1$, the sum of the eigenvalues of the contingency table, including the trivial component, and the absolute value of the correlation ratio, which is 1. These statistics, however, have not been further examined or used, and their merits for helping the interpretation of results are not known.

Of these, Kalantari, Lari, Rizzi and Simeone (1993) raised a question about Cramér's upperbound, stating that it can rarely be attained, and proposed another upper bound, calculated by complex computations. If we use their upper bound, we will obtain yet another statistic. Interested readers are referred to their paper.

6.4 Is My Pet a Flagrant Biter?

This example was reported by Sheskin (1997), and with the kind permission of the publisher CRC the data table is reproduced here. Sixty mice, 50 gerbils, 90 hamsters and 80 guinea pigs were observed over two weeks and categorized in one of the following categories: (1) not a biter, (2) mild biter, (3) flagrant biter. The data are listed in Table 6.5. Sheskin (1997) used this example to explain chi-square analysis. Our analysis

Table 6.6 *Order-0 Approximation*

	Not a biter	Mild Biter	Flagrant Biter
	(No)	(Mild)	(Biter)
Mice	25.5	14.4	20.1
Gerbils	21.3	12.0	16.8
Hamsters	38.3	21.5	30.2
Guinea Pigs	34.0	19.1	26.9

Table 6.7 *Residual: Data minus Order-0 Approximation*

	No	Mild	Biter
Mice	-5.5	1.6	3.9
Gerbils	8.8	-2.0	-6.8
Hamsters	11.8	8.5	-20.2
Guinea Pigs	-15.0	-8.1	23.1

can be regarded as multidimensional analysis of the over-all chi-square statistic. For this table, the chi-square is 59.16 with 6 degrees of freedom, which is significant: rows and columns are not statistically independent. Since the table is 4×3, this significant association between rows and columns can be exhaustively explained by two components. Since this is a small example, let us look at our quantification analysis in detail (Tables 6.7-6.12).

We can tell that the first component is very dominant, explaining 94

Table 6.8 *Order-1 Approximation and Residual*

	No	Mild	Biter	No	Mild	Biter
Mice	22.7	13.1	24.2	-2.7	2.9	-0.2
Gerbils	26.1	14.2	9.7	3.9	-4.2	0.3
Hamsters	52.1	27.8	10.2	-2.1	2.2	-0.2
Guinea Pigs	18.1	12.0	49.9	0.9	-1.0	0.1

Table 6.9 *Order-2 Approximation and Residual*

	No	Mild	Biter	No	Mild	Biter
Mice	20.0	16.0	24.0	0.0	0.0	0.0
Gerbils	30.0	10.0	10.0	0.0	0.0	0.0
Hamsters	50.0	30.0	10.0	0.0	0.0	0.0
Guinea Pigs	19.0	11.0	50.0	0.0	0.0	0.0

Table 6.10 *Summary Statistics*

Component	1	2
Eigenvalue	.20	.01
Singular Value	.45	.11
Delta	94.2	5.8
Cum.Delta	94.2	100.00

Table 6.11 *Normed and Projected Weights for Pets*

Weight	y_1	y_2	$\rho_1 y_1$	$\rho_2 y_2$
Mice	.32	1.09	.14	.12
Gerbils	-.67	-1.87	-.30	-.21
Hamsters	-1.06	.56	-.47	.06
Guinea Pigs	1.37	-.28	.61	-.03

Table 6.12 *Normed and Projected Weights for Pets*

Weight	x_1	x_2	$\rho_1 x_1$	$\rho_2 x_2$
No	-.77	-.88	-.34	-.10
Mild	-.61	1.68	-.27	.19
Bite	1.40	-.09	.63	-.01

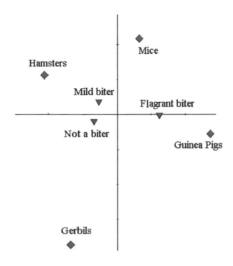

Figure 6.1 *Biting Behavior Projected onto Pets*

percent of the total variance. This dominance of the first component is typically used to justify an interpretation of the results only in terms of the first component. This practice is generally accepted, but we should at least know what it means. The case in point is about its geometry. As mentioned briefly in Chapter 2, the row space and the column space are not the same unless the singular value is 1, meaning that we must be careful in plotting both rows and columns in the same graph, that is, a joint plot. Geometrically correct joint plots, therefore, are to plot normed row weights and projected column weights (i.e., we project column weights onto row space), and to plot projected row weights and normed column weights (i.e., projecting row weights onto column space). Empirically we know, however, that such plots are difficult to interpret because projected weights have typically much smaller dispersions than normed weights. See Figures 6.1 and 6.2.

These graphs may be interpretable at a first glance, but we should note that the graph with normed weights is used for convenience and

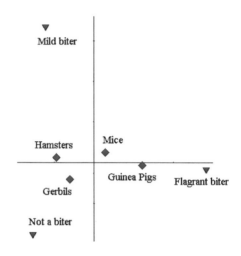

Figure 6.2 *Pets Projected onto Biting Behavior*

that it does not contain a crucial piece of information, that is, singular value. Thus, although geometrically correct, those graphs in Figures 6.1 and 6.2 are not used in practice for analysis of incidence data. The most popular choice of graphical display is what we call symmetric scaling, sometimes referred to as the French method. It is a joint plot of both projected row weights and projected column weights, the only problem being that the row space and the column space are generally different. This last point indicates that we are overlaying two planes in the case of a two-dimensional graph. Let us look at the graph using symmetric scaling (Figure 6.3).

Let us examine this symmetric joint graph very closely. We often hear about ignoring the second and the remaining components because δ_1 is very high, that is, 94 percent. As pointed out by Nishisato (1996), δ is a difficult statistic to interpret, in particular when we deal with incidence data. We must also look at ρ_1. In the current example, δ_1 is 94 percent, and ρ_1 is 0.45. Notice that in spite of a very high value of

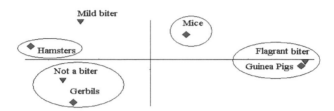

Figure 6.3 *Symmetric Scaling of Biting Behavior and Pets*

δ, the singular value is small, indicating that the discrepancy between row space and column space is large. In terms of the angle between the row continuum and the column continuum, the discrepancy is 63 degrees, that is, $\cos^{-1} \rho_1 = 63.48$. This is a very large angle separation, for in most instances we go ahead with this space discrepancy and interpret the joint plot with some cautions as given by Lebart, Morineau and Warwick (1984). How about the discrepancy between the row continuum and the column continuum in the second dimension? It is 84 degrees! This is obtained by $\cos^{-1} \rho_2$, where ρ_2 is 0.11.

With this much knowledge, let us look at the symmetric graph again. The cautionary remark by Lebart, Morineau and Warwick (1984) now rests heavily on us. It is not a matter of overlaying two planes with one being slightly slanted, but the slant is complicated because of involvement of two components. As suggested by Nishisato and Clavel (2002), we actually need three-dimensional space to accommodate relations created by two components, one additional dimension due to the discrepancy of two spaces. The only encouraging aspect of the current problem is that the contribution of the second component is so small that the space discrepancy would not place the data points much further away from where they are now: Imagine moving a point with the current coordinate around the horizontal axis, while maintaining the distance of the point from the horizontal axis, that is, rotating the point around the horizontal axis. See that it will not move away very much from where it is now.

Though the space discrepancy is the main problem with symmetric scaling, the graph nevertheless presents us a general grasp of relationships between row categories (pets) and column categories (biting behavior). Figure 6.3 is a graph which is correct if the space discrepancy does not exist: In this case, guinea pigs are flagrant biters, gerbils non-biters, hamsters between non-biters and mild biters and mice between flagrant biters and mild biters. However, if we consider ignoring the second component for the time being and rotate the pet axis by 63 degrees, the situation changes slightly by taking guinea pigs from flagrant biters, and gerbils and hamsters away from mild biters and non-biters. At this stage, we then introduce the second component by ignoring its space discrepancy because the variance is comparatively small. Then, very roughly though, we arrive at a similar conclusion that we draw from the symmetric scaling: guinea pigs are closest to flagrant biters, gerbils closest to non-biters, hamsters closest to mild biters and mice still between flagrant biters and mild biters. But, we must admit that these associations between pets and biting behavior are as not clear cut as one may speculate from the symmetric scaling, and we should realize how important to listen to the cautions by Lebart, Morineau and Warwick (1984). Out of the three modes of joint plots, we will adopt the symmetric scaling for analysis of incidence data. We will see later that this problem of incidence data is quite different from that of dominance data.

6.5 Supplementary Notes

One may wonder if the decomposition of the total χ^2 with the degrees of freedom $(m - 1)(n - 1)$ could be incorporated for testing the significance of individual components. Nishisato (1980a) traced a number of developments and adopted a combination of Bartlett's approximation (1947), Bock's modification (1960) and Rao's decomposition of the degrees of freedom (1952), and presented the following formula for testing the significance of Component k,

$$\chi^2(k) = -[f_t - 1 - \frac{1}{2}(m+n)]\log_e(1 - \rho_k^2), \ df = m+n-2k \ (6.11)$$

Notice the following,

$$\sum_{k=1}^{K} \chi^2(k) = -[f_t-1-\frac{1}{2}(m+n)] \sum_{k=1}^{K} \log_e(1-\rho_k{}^2), \ df = (m-1)(n-1)$$

(6.12)

The literature suggests that $\chi^2(1)$ tends to over-estimate the chi-square and that $\chi^2(K)$ under-estimates it. In addition to the distributional problem, criticisms were raised on the above integration of pieces of the results into a single statistic. Thus, the above statistic should be used only to obtain a rough indication of the substance of each component.

As for the chi-square statistic for contingency tables, it is well known that Cochran (1954) investigated what happens when the expected frequency falls below a certain number such as 5, and that Craddock and Flood (1970) investigated what happens to the chi-square approximation when the table becomes small, for instance, from 5x5 to 3x2. In these cases, the chi-square approximation may become unsatisfactory, another problem we must be concerned with.

CHAPTER 7

Multiple-Choice Data

7.1 Example

Multiple-choice data consist of a table of N rows (subjects) by n columns (items) of chosen response option numbers. The following is an example of multiple-choice data collected from 3 subjects on 5 questions about food preferences and personality. The questions are:

1. Which do you like best, (1=beer, 2= whiskey, 3=wine)?
2. Do you say things that you will later regret, (1=never, 2=sometimes, 3=often)?
3. How would you describe yourself, (1=outgoing, 2=homebody)?
4. Do you have migraines, (1=never, 2=occasionally, 3=often)?
5. Do you like milk, (1=yes, 2=no)?

Table 7.1 *Multiple-Choice Data*

Subject	Item 1	Item 2	Item 3	Item 4	Item 5
1	1	3	1	3	2
2	3	2	2	1	1
3	2	1	1	2	2

The elements of this data table indicate chosen response options. The five questions have 3, 3, 2, 3, 2 response options, respectively. Each column contains distinct numbers, 1, 2, or 3 for the first four columns (questions) and 1 or 2, for the last question.

The above form of multiple-choice data is only a practical way of describing the data. For quantification, the data are transformed into a table of (1,0) response patterns, where "1" means a response (choice of the option) and "0" absence of a response. Thus, if the above sample data are to be subjected to quantification, the table is transformed to the

Table 7.2 *Response-Pattern Table*

Subject	1	2	3	4	5
1	100	001	10	001	01
2	001	010	01	100	10
3	010	100	10	010	01

response-pattern table as in Table 7.2. Let N be the number of subjects and m_j be the number of response options of item j, then the data matrix for item j is expressed as an $N \times m_j$ matrix $\mathbf{F}_j$ such that each of N rows would look like $(0,1,0,...,0)$, containing only one choice ("1") out of m_j options, and the entire data matrix for N subjects and n items can be expressed as:

$$\mathbf{F} = [\mathbf{F}_1, \mathbf{F}_2, \mathbf{F}_3, \ldots, \mathbf{F}_n] \tag{7.1}$$

In the above example,

$$\mathbf{F}_1 = \begin{bmatrix} 1 & 0 & 0 \\ 0 & 0 & 1 \\ 0 & 1 & 0 \end{bmatrix} \tag{7.2}$$

Let m indicate the total number of options of n items, that is,

$$m = \sum_{j=1}^{n} m_j \tag{7.3}$$

The task of quantification is to determine m option weights in such a way that, for example, the variance of subjects' weighted scores be a maximum, subject to the constraint on the unit of measurement.

7.2 Early Work

Guttman (1941) presented a detailed and thorough treatment of the quantification problem. He provided three distinct formulations for quantifying multiple-choice data:

1. Determine scores for subjects so as to maximize the between-option sum of squares, relative to the total sum of squares

2. Determine weights for options so as to maximize the between-subject sum of squares, relative to the total sum of squares

3. Determine scores of subjects and weights for options so as to

maximize the product-moment correlation of data weighted by subjects' scores and those weighted by option weights.

Guttman showed that these three procedures are mathematically equivalent as did Maung (1941a) for the contingency table.

It is of historical interest that the pioneers in quantification of multiple-choice data adopted the above way of placing response-pattern matrices in a row. A direct extension of the contingency formulation would have been to consider an n-dimensional contingency table and its associated decomposition formula. For example, for $n=3$, that is, a 3-dimensional contingency table, we would have the following formula,

$$f_{ijk} = \frac{f_{i..}f_{.j.}f_{..k}}{f_{...}}[1 + \rho_1 y_{1i}x_{1j}z_{1k} + \rho_2 y_{2i}x_{2j}z_{2k} \cdots + \rho_t y_{ti}x_{tj}z_{tk}]$$

(7.4)

While it is possible to extend this expression to an n-dimensional contingency table, it would be extremely difficult to derive a practical solution to it, not to mention the interpretation of the outcome. In addition, it is quite likely that a multidimensional contingency table would have many empty cells, due to a sheer number of combinations of options. For some reason or other, Guttman and the followers opted for the simpler way of dealing with a string of response-pattern matrices. In contrast, much later, the decomposition of a multi-dimensional contingency table was realized and gained the popularity in studies on log linear analysis. Notice, however, that we encounter the same number problem with log linear analysis, too, and we have the same limitation to the number of variables and that of categories which we can handle in practice.

7.3 Some Basics

Consider n multiple-choice items, with item j having m_j options. Each of N subjects is asked to choose one option per item, that is, $\mathbf{F}_j\mathbf{1} = \mathbf{1}$.

As is the case with the contingency table, we impose the condition that the sum of all the weighted responses is zero, that is, $\mathbf{1}'\mathbf{Fx} = 0$. Guttman (1941) noted an interesting aspect of this condition: it means that the sum of weighted responses of each item is zero, too, that is,

$$\mathbf{f}'\mathbf{x} = 0 \Rightarrow \mathbf{f}_j'\mathbf{x}_j = 0, j = 1, 2, \cdots, n. \tag{7.5}$$

Some investigators wonder why the means of all items should be set equal to zero, for some items are more relevant than others to the measurement under consideration. However, Nishisato (1980a) pointed out an interesting merit of the condition: Consider two dichotomous questions in a mathematics test, one of which is a very easy question such that some 90 percent of people would answer it correctly, the other of which is a very difficult question such that only 10 percent would answer it correctly. Then, the above condition of zero origin for both items would yield such a scoring scheme that if you miss an easy question, the penalty is great; if you answer a difficult question correctly, the reward is great. For example, weights for these questions can be -4.5 for the incorrect answer and 0.5 for the correct answer to the easy item (i.e., -4.5x10+0.5x90 =0), and -0.5 for the incorrect answer and 4.5 for the correct answer to the difficult item (i.e., -0.5x90+4.5x10=0). Thus, if you miss an easy item, the penalty is large (-4.5) and if you miss a difficult item, the penalty is not great (-0.5). The condition of equal average contributions of items actually takes care of the effect of the so-called difficulty factor.

Multiple-choice data also have a trivial solution. Thus, $T(comp)$ is affected by this condition. Assuming that the number of subjects is larger than the total number of response options m, the rank of the response-pattern matrix is in general given by

$$rank(\mathbf{F}) = rank[\mathbf{F}_1, \mathbf{F}_2, \cdots, \mathbf{F}_n] = \sum_{j=1}^{n}(m_j - 1) + 1 = m - n + 1$$

$$(7.6)$$

Since one trivial solution is involved and it is removed from analysis,

$$T(comp) = m - n \qquad (7.7)$$

If the number of subjects is smaller than $(m - n)$, then

$$T(comp) = N - 1 \qquad (7.8)$$

The total information, $T(inf)$, is given (Nishisato, 1994) by:

$$T(inf) = \sum_{k=1}^{m-n} \rho_k^2 = \frac{\sum_{j=1}^{n} m_j}{n} - 1 = \frac{m}{n} - 1 = \overline{m} - 1 \qquad (7.9)$$

where $\overline{m}$ is the average number of options. The definition of δ_k is the same as the contingency table, that is, the percentage of the eigenvalue or squared singular value divided by the sum of all eigenvalues, which can be expressed as

$$\delta_j = \frac{100\rho_j^2}{\overline{m} - 1} \qquad (7.10)$$

Thus, $T(inf)$ is a function of the number of options (categories).

Many key statistics associated with quantification of multiple-choice data are related to the number of categories of variables. For example, the total number of dimensions required to accommodate a variable with m_j categories is

$$N_j(dim) = m_j - 1 \qquad (7.11)$$

The total number of dimensions needed for n variables is

$$N_T(dim) = \sum_{j=1}^{n}(m_j - 1) = \sum_{j=1}^{n} m_j - n = m - n \qquad (7.12)$$

In 1958, Lord demonstrated that the principal component scores maximize the generalized Kuder-Richardson reliability coefficient, or more popularly Cronbach's α. Indeed, our quantification procedure yields scores for subjects with a maximal value of α. This is obvious from the following relation (Nishisato, 1980a):

$$\alpha = 1 - \frac{1 - \rho^2}{(n-1)\rho^2} \qquad (7.13)$$

It is known (Nishisato, 1980a, 1994) that the average information in multiple-choice data is

$$\overline{\rho}^2 = \frac{\sum_{k=1}^{m-n} \rho_k^2}{m - n} = \frac{\overline{m} - 1}{m - n} = \frac{1}{n} \qquad (7.14)$$

and that α becomes negative when ρ^2 of the component gets smaller than the average information.

The concept of reliability is defined as the ratio of the variance of the 'true scores' to the variance of the observed scores, and as such the reliability is a positive quantity. Nevertheless, the above discussion indicates that the coefficients calculated by the formula for α are negative whenever the eigenvalue becomes smaller than the average eigenvalue for a given data set. In this case, negative coefficients of α can be observed in nearly half the components. How should we interpret a negative coefficient of reliability? If we see the definition of reliability, there is no rational way to interpret it. Therefore, Nishisato (1980a) suggested stopping the extraction of components as soon as ρ^2 becomes smaller than $1/n$. Accordingly, we introduce an adjusted statistic δ^*_k as the percentage of ρ_k^2 over the sum of ρ_j^2 greater than $1/n$,

$$\delta^*_k = \frac{100\rho_k^2}{\sum_{(\rho_j^2 \text{greater than} \frac{1}{n})}^{m-n} \rho_j^2} \qquad (7.15)$$

For multiple-choice data, most researchers would notice that δ is generally unexpectedly low even for the first component. Benzécri (1979) discussed the matter, and it is interesting to note his statement that components of interest are those of eigenvalues greater than $\frac{1}{n}$.

α is not an easy statistic to interpret, for, first of all, it is a function of the number of items: if the number of items increases in a test, so does α, and secondly its sampling distribution is very complex and cannot be interpreted in the same way as product-moment correlation. In practice, researchers would wish to tell whether the reliability of a test is good enough for using it in practice. Or, more commonly, when can we say that the reliability of a test is poor, good or excellent, or acceptable?

One way to look at α is to calculate the effective test length derived from the Spearman-Brown formula (e.g., see Nishisato, 1980a). Let α_o be the reliability coefficient of the original test and α the reliability of the test lengthened h times by statistically equivalent items. The Spearman-Brown formula then gives the reliability of the lengthened test as

$$\alpha = \frac{h\alpha_c}{1 + (h - 1)\alpha_c} \tag{7.16}$$

We can rewrite this formula as

$$h = \frac{\alpha(1 - \alpha_c)}{\alpha_c(1 - \alpha)} \tag{7.17}$$

where h is called the effective test length. We can use this statistic h in the following way. Suppose we obtain α of 0.67 from 10-item data subjectively scored and α of 0.89 from the same 10-item data quantified by MUNDA. Then, let α_o equal to 0.67 and α to 0.89, which results in h=3.99, that is approximately 4. This means that if we wish to increase the reliability of subjectively scored data to 0.89, we must increase the number of similar items in the test by four times the current number, that is, 40-item test from 10 items. In this way, we can show the impact of scaling more clearly than compare two coefficients of reliability.

Let us look at the example of multiple-choice data, given at the beginning of this chapter. Suppose we assign weights $x_1, x_2, ..., x_{13}$ to the 13 options (3+3+2+3+2 options for items 1, 2, 3, 4 and 5, respectively) and express the original data in terms of weighted responses (Table 7.3).

These are called item scores (j=1, 2, 3, 4, 5) and the total score (t). The correlation between each item score and the total score is called *item-total correlation*, and for item j it is indicated by r_{jt}. In matrix notation,

Table 7.3 *Multiple-Choice Data*

Subject	Item 1	Item 2	Item 3	Item 4	Item 5
1	x_1	x_6	x_7	x_{11}	x_{13}
2	x_3	x_5	x_8	x_9	x_{12}
3	x_2	x_4	x_7	x_{10}	x_{13}

Subject	Total Score (t)
1	$x_1 + x_6 + x_7 + x_{11} + x_{13}$
2	$x_3 + x_5 + x_8 + x_9 + x_{12}$
3	$x_2 + x_4 + x_7 + x_{10} + x_{13}$

it can be expressed as

$$r_{jt} = \frac{\mathbf{x}'_j \mathbf{F}'_j \mathbf{F} \mathbf{x}}{\sqrt{\mathbf{x}'_j \mathbf{D}_j \mathbf{x}_j \mathbf{x}' \mathbf{F}' \mathbf{F} \mathbf{x}}} \qquad (7.18)$$

where $\mathbf{x}_j$ is the vector of weights for options of item j, $\mathbf{F}_j$ is the $N \times m_j$ matrix of response patterns for item j, $\mathbf{D}_J$ is the diagonal matrix of the column totals of $\mathbf{F}_j$.

There are a number of interesting roles that the item-total correlation, or rather the squared item total correlation, can play in quantification (Nishisato, 1994). Some of the examples are:

(1) The sum of squares of scores of item j is proportional to the squared item-total correlation,

$$SS(j) = \mathbf{x}'_j \mathbf{D}_j \mathbf{x}_j \propto r_{jt}^2 \qquad (7.19)$$

where the sign '$\propto$' indicates 'proportional to'. This is extremely important to note, for it means that when the item is highly relevant to the measurement (i.e., r_{jt} is high), high scorers tend to choose highly weighted options and low scorers negatively weighted options. When the item is irrelevant to the measurement by the test, it does not make any difference which option one may choose, for most options have weights close to zero, and when the item is highly relevant, it matters which option one may choose, for the weights of options then are widely distributed and one's total score is strongly influenced by which option of the item one chooses.

(2) The eigenvalue (the squared singular value, the correlation ratio) is the average of the squared item-total correlation coefficients of all the items.

$$\rho_k{}^2 = \frac{\sum_{j=1}^{n} r_{jt}{}^2}{n} \tag{7.20}$$

This is a useful relation to remember.

(3) Using the above relation, we can rewrite the formula for the reliability coefficient α,

$$\alpha = 1 - \frac{1 - \frac{\sum r_{jt}{}^2}{n}}{(n-1)\frac{\sum r_{jt}{}^2}{n}} = \frac{n}{n-1}\left(\frac{\sum r_{jt}{}^2 - 1}{\sum r_{jt}{}^2}\right) \tag{7.21}$$

This formula tells us how important the squared item-total correlation is for the reliability of scores, and it suggests one way to increase the reliability, that is, by collecting items which have high item-total correlation coefficients. In particular, if all items are perfectly correlated with the total score, then the sum of the squared item-total correlations is equal to n. Thus, if you substitute this for the sums in the above equation, we obtain

$$\alpha = \frac{n}{n-1}\left(\frac{\sum r_{jt}{}^2 - 1}{\sum r_{jt}{}^2}\right) = \frac{n}{n-1}\left(\frac{n-1}{n}\right) = 1 \tag{7.22}$$

As for reliability itself, we have already noted that our quantification method determines option weights so as to yield scores with a maximum value of the generalized Kuder-Richardson internal consistency reliability, or the Cronbach α (Lord, 1958).

The above formula also shows that α becomes zero when the sum of squared item total correlations is 1, and that α becomes negative when this sum becomes less than 1.

(4) The total contribution of a variable with m_j categories can be expressed as

$$\sum_{k=1}^{m-n} r_{jt(k)}^2 = m_j - 1 \tag{7.23}$$

In practice, we often encounter a multiple-choice questionnaire with different numbers of response options for different items. To the extent possible, we should try to offer the same number of response options to all the items. To see this point from another angle, consider the sum of squares of option p of item j, SS(jp), and the sum of squares of item j, SS(j), which are given (Nishisato, 1996) by

$$SS(jp) = \sum_{k=1}^{m-n} f_{jp}x_{jpk}{}^2 = n(N - f_{jp}) \qquad (7.24)$$

where f_{jp} is the frequency of option p of item j.

$$SS(j) = \sum_{j}^{m_j} SS(jp) = \sum_{k=1}^{m-n} \sum_{p=1}^{m_j} f_{jp}x_{jpq}^2 = nN(m_j - 1) \qquad (7.25)$$

The first formula shows clearly that the smaller the number of people who choose option p, the greater the option contribution to the total space, contrary to what one might wish to see. The second formula shows that the more options we have the greater the contribution of the item to the total space. These observations later will be demonstrated by a numerical example and we will see how "outliers" are born in the process of quantification. When items are dichotomous, Yamada and Nishisato (1993) provided a number of interesting properties of dual scaling.

7.4 Future Use of English by Students in Hong Kong

Peter Tung, who is currently a professor of education in Hong Kong, collected data for Nishisato's class on scaling as a part of the course assignment. His original questionnaire had 11 items, and he collected data from 50 high school students. Analysis of his complete data were reported in Nishisato (1994). Since some of the questions are of no interest to our discussion, we will look at only a part of the original data, that is, responses from 50 students on five questions, instead of 11 questions, with the kind permission of Peter Tung.

  This data set offers an interesting problem of what we may call an "outlier" problem. Typically, an outlier in MUNDA is interpreted to mean a response that exerts an exceeding influence on the outcome of data analysis. Before discussing this outlier problem, let us look at the distribution of information over components.

  Notice that the eigenvalue of component 6 is 0.19, which is smaller than 1/5, that is, 0.20, resulting in a negative value of the reliability coefficient α. Thus, we consider only the first five components, and adjusted delta values are therefore listed in the last row of Table 7.6. The first two components account for more than 50 percent of the total information,

Table 7.4 *Tung's Five Questions*

Item	Question "**After graduation from high school,**
1	Will you use English to talk to others? 1=most likely; 2=very likely; 3=likely; 4=unlikely; 5=I don't know
2	Will you use English to talk to others at home? 1=yes; 2=no; 3=I don't know
3	Will you use English to talk with others at school gatherings? 1=yes; 2=no; 3=I don't know
4	Will you use English to talk with others at your job? 1=yes; 2=no; 3=I don't know
5	Will you use English to talk with others in pursuing your future studies? 1=yes; 2=no; 3=I don't know

and the joint plot is as shown in Figure 7.1. Subjects are indicated by solid squares and options by empty inverted triangles.

We see a strange distribution of subjects and response options. A dense cluster of data points lies along Component 2 (vertical axis). Component 1 (horizontal axis) has a wider scatter than Component 2, as we expect, but the strange aspect of this is that this wider scatter of Component 1 is represented by only a very few data points. In other words, Component 1 is determined by only a few data points. This is a typical phenomenon of the so-called outliers in the data set.

Let us examine the matter a little deeper. In Figure 7.1, the main key data points are labeled, namely, Option 4 of Item 1, Option 2 of Item 4, Subject 20, Option 2 of Item 3, and Subject 4. Since Figure 7.1 is a joint plot of symmetric scaling, let us look at projected weights of subjects and options of items of Component 1.

When we pick up outstandingly large scores and weights and look at the original data in Table 7.5, we discover the following:

(1) Only Subject 20 who obtained the score of 4.90 chose Option 4 of Item 1 and Option 2 of Item 4, both of which have the same weight of 6.73.

(2) Subject 20 and Subject 4 are the only subjects who chose Option 2 of Item 3, which has the weight of 4.17.

The above points indicate that optimal weights and scores are

Table 7.5 *Peter Tung's Data*

Ss	Q1	Q2	Q3	Q4	Q5	Ss	Q1	Q2	Q3	Q4	Q5
1	1	2	1	1	1	26	1	3	1	1	1
2	1	1	1	1	1	27	2	2	1	1	1
3	1	2	1	1	1	28	1	1	1	1	1
4	2	2	2	1	1	29	1	3	1	1	1
5	1	2	1	1	1	30	5	2	3	3	1
6	2	2	3	1	1	31	1	1	1	1	1
7	1	2	1	1	1	32	2	3	3	1	1
8	1	3	3	1	1	33	1	3	1	1	1
9	2	2	1	1	1	34	1	3	1	1	1
10	2	2	1	1	1	35	1	1	1	1	1
11	1	2	3	1	1	36	1	2	1	1	1
12	2	1	1	1	1	37	1	1	1	1	1
13	2	3	1	1	1	38	1	3	1	1	1
14	3	2	3	1	1	39	1	3	1	1	1
15	2	2	3	3	1	40	2	2	1	1	1
16	1	2	3	1	1	41	3	2	1	1	1
17	3	1	3	3	2	42	2	2	1	3	1
18	1	2	1	1	1	43	3	2	1	1	1
19	1	2	1	1	1	44	2	2	1	1	1
20	4	2	2	2	1	45	1	1	1	3	1
21	3	2	1	1	1	46	1	2	1	1	1
22	3	2	1	1	1	47	1	1	1	1	1
23	5	2	1	1	2	48	1	3	1	1	1
24	2	2	1	1	1	49	3	2	1	1	1
25	2	1	1	1	1	50	2	2	1	1	1

Table 7.6 *Distribution of Information*

Component	1	2	3	4	5	6
Eigenvalue	.53	.40	.26	.22	.21	.19
Singular Value	.73	.63	.51	.47	.46	.44
Reliability	.78	.63	.29	.13	.06	-.06
Delta	24.1	18.2	11.8	10.2	9.5	8.7
Cum.Delta	24.1	42.4	54.1	64.3	73.8	82.5
Adjusted Delta	32.7	24.7	16.1	13.6	13.0	

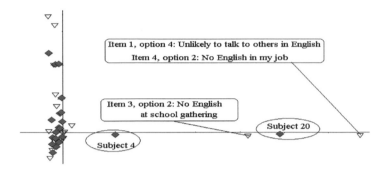

Figure 7.1 *Joint Plot of the First Two Components*

reciprocally related to the frequencies. Recall the formula for the sum of Option p of Item j, $SS(jp)$, mentioned earlier, that is,

$$SS(jp) = \sum_{k=1}^{m-n} f_{jp} x^2{}_{jpk} = n(N - f_{jp})$$

where N is the number of subjects, f_{jp} the frequency of Option p of Item j, n the number of items, and m the total number of options. Thus the smaller the number of subjects who choose Option p the greater the contribution of the option. In the current example, $SS(14)=SS(42)=$ $5(50-1)=245$. This is the contribution in terms of the sum of squares of quantified responses of these options to the total space. Compare this, for example, with the total sum of squares of Option 1 of Item 1, which is, $5(50-44)=30$. This comparison clarifies why those options with small frequencies are likely to dominate some components. Recall further that our method seeks components with a maximal sum of squares.

The above observation can also be reached from a different point of view. As we discussed, the sum of weighted responses of each item is set to zero and our method deals with the chi-square metric. Thus, if an item has two options and the response frequencies of the two options are, for instance, 1 and 99, the corresponding weights are proportional to -99 for the first option and 1 for the second option. This means that an option with a small frequency is located far from the origin and an option with

Table 7.7 *Projected Weights of Subjects and Options of Component 1*

Ss	Score	Ss	Score	Item	Option	Weight
1	-.1	26	-.2	1	1	-.2
2	-.2	27	-.0		2	.0
3	-.1	28	-.2		3	-.1
4	1.2	29	-.2		4	6.7
5	-.1	30	-.1		5	-.2
6	-.0	31	-.2			
7	-.1	32	-.1	2	1	-.3
8	-.2	33	-.2		2	.2
9	-.0	34	-.2		3	-.3
10	-.0	35	-.2			
11	-.1	36	-.1	3	1	-.2
12	-.2	37	-.2		2	4.2
13	-.2	38	-.2		3	-.2
14	-.1	39	-.2			
15	-.0	40	-.0	4	1	-.1
16	-.1	41	-.1		2	6.7
17	-.3	42	-.0		3	-.2
18	-.1	43	-.1			
19	-.1	44	-.0	5	1	.0
20	4.9	45	-.3		2	-.3
21	-.1	46	-.1		3	.0
22	-.1	47	-.2			
23	-.2	48	-.2			
24	-.0	49	-.1			
25	-.2	50	-.0			

a large frequency is located near the origin. In the current example, it so happened that one subject (Subject 20) chose those options of very small frequencies (Option 4 of Item 1 and Option 2 of Item 4), which led to a very high score for the subject.

This may be an appropriate time to draw our attention to the issue of the number of options per item. As mentioned earlier, the sum of squares of Item j, $SS(j)$, is proportional to the number of options of the item,

$$SS(j) = \sum_{k=1}^{m-n} \sum_{p=1}^{m_j} f_{jp} x^2_{jpk} = nN(m_j - 1)$$

Therefore, it is not advisable to introduce an excessively large number of

Table 7.8 *Inter-Item Correlation for Component 1 (250 Responses)*

Item	1	2	3	4	5
1	1.00				
2	.15	1.00			
3	**.71**	.17	1.00		
4	**.99**	.12	**.70**	1.00	
5	.03	.05	.04	.03	1.00

options (e.g., 12) to the data set where most items have a smaller number of options (e.g., 3). It is obvious what may happen if one item has 12 options and 9 other items have three options each: the 12-option item is likely to dominate in many components. Thus, it is important to equate contributions of items to the total information by designing an instrument with all questions having the same number of options.

The number of options also has an implication for what degree of a nonlinear relation it can capture. If the number of options is two, we cannot capture nonlinear relations; if the number is three, we can capture linear and quadratic relations; with four options, we can capture linear, quadratic and quartic relations, and so on. Thus, the number of options minus one is the degree of a polynomial relation, up to which we can capture nonlinear relations.

We see a conflict here regarding the number of options. We need an appropriate number of options to capture a variety of nonlinear relations, and at the same time, an increase in the number of options may create too much information to analyze in practice. Since the latter is a more crucial problem than the former, the general consensus seems to use the same number of options for all questions and the number should be between 3 and 5.

Let us get back to the results of analysis. We have seen that the first component seems to reflect the influence of a few responses, namely the responses to Item 1, Item 4 and Item 3 of Subject 20 and the response to Item 3 of Subject 4, which we may call outliers in the data set. What is startling is that these four responses alone could have created the first component. To see this possibility, let us look at the inter-item correlation matrix associated with the first component (Table 7.8). Notice that high inter-item correlation coefficients in bold face are all associated with Items 1,3 and 4, which involve outlier responses. Let us carry out

Table 7.9 *Inter-Item Correlation for Component 1 (246 Responses)*

Item	1	2	3	4	5
1	1.00				
2	.29	1.00			
3	**.23**	.01	1.00		
4	**.30**	.14	**.36**	1.00	
5	.52	.07	.17	.27	1.00

Table 7.10 *Inter-Item Correlation for Component 2*

Item	1	2	3	4	5
1	1.00				
2	.29	1.00			
3	**.23**	.00	1.00		
4	**.30**	.14	**.36**	1.00	
5	.52	.08	.17	.27	1.00

a little experiment: Suppose that we replace responses of Subject 20 to Items 1, 3 and 4 with zeros and the response of Subject 4 to Item 3 with a zero, that is, we treat these responses as missing. What will happen if we ignore these four responses out of 250 (i.e., 50x5) responses? The inter-item correlation matrix associated with the first component of 246 responses changes to Table 7.9. Notice those bold-faced correlation co-efficients. We ignored only four responses out of 250 responses (=50x5). Yet, what a remarkable change in the correlation matrix! Let us then move on to look at the inter-item correlation matrix of the original data associated with the second component (Table 7.10). These correlations are almost identical to those in Table 7.9, the only difference being the correlation between Item 2 and Item 5 (0.07 in Table 7.9, 0.08 in Table 7.10). In other words, the correlation matrix associated with the second component of the original data is essentially identical to that of the first component of the data without those four responses. Therefore, we can clearly conclude that the first component of the original data was created by those four outlier responses.

The first component in Figure 7.1, therefore, depicts the outliers,

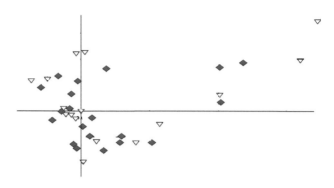

Figure 7.2 *Joint Plot of Components 2 and 3 of Complete Data*

Table 7.11 *Distribution of Information (Complete Data)*

Component	1	2	3	4	5	6
Eigenvalue	.53	.40	.26	.23	.21	.19
Singular Value	.73	.63	.51	.47	.46	.44
Reliability	.78	.63	.29	.13	.06	-.06

namely Subject 20 and his/her responses to Items 1, 3 and 4 and Subject 4 and his/her response to Item 3. Then, it follows that the symmetric joint plot of Component 2 and Component 3 of the original complete data (see Figure 7.2) should be very similar to that of Component 1 and Component 2 of the data without those outlier responses, that is, the reduced data (see Figure 7.3). The two graphs seem to verify the above conjecture. Notice, of course, the numbers of data points in two graphs are different, hence resulting necessarily in minor differences between them.

That the original data contained four outlier responses can also be viewed from the comparison of information over components of the original complete data with the reduced data (Tables 7.11 and 7.12). Notice a

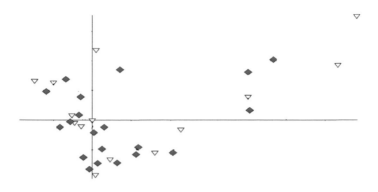

Figure 7.3 *Joint Plot of Components 1 and 2 of Reduced Data*

Table 7.12 *Distribution of Information (Reduced Data)*

Component	1	2	3	4	5
Eigenvalue	.40	.27	.22	.21	.19
Singular Value	.63	.52	.47	.46	.44
Reliability	.63	.32	.13	.06	-.06

remarkable resemblance between the two tables. We can thus call the first component of our complete data an outlier component.

This data set has more aspects we should look at: Subjects 23 and 30 are the only ones who chose Option 5 of Item 1; Subjects 17 and 23 are the only ones who chose Option 2 of Item 5. Thus, this data set is marred by the conditions that a few categories of very small frequencies surface as dominant powers in determining the distribution of information. These unpopular options are negative responses that they do not use English in talking to others and in future studies after graduation from high school. Once these negative options are removed, however, there remains little information worth analyzing or looking at.

The current example presents an excellent opportunity to see what happens when the distribution of responses over response options is much

Table 7.13 *Information Distribution over Five Components*

Component	1	2	3	4	5
Eigenvalue	.54	.37	.35	.31	.13
Singular Value	.74	.61	.59	.55	.36
Alpha	.83	.67	.62	.55	-0.33
Delta	27.2	18.7	17.3	15.4	6.5
Cum. Delta	27.2	45.9	63.2	78.6	85.1
Adjusted Delta	34.4	23.6	22.3	19.8	

skewed. Is this a design problem for researchers? No, it so happened that the problem under investigation was directed to a rare phenomenon. Peter Tung's original data contained responses to 11 questions, rather than five used here. We employed a subset of the original data to make a point that outlier responses or skewed distributions of responses can destroy data analysis.

7.5 Blood Pressures, Migraines and Age Revisited

We have discussed the analysis of a small data set on blood pressure and others in Chapter 1 to stress the need for nonlinear analysis. The questionnaire and data were presented in Table 1.8 and Table 1.9, respectively. Let us present a more detailed analysis here. Since there are 15 subjects, six questions with three response options per question, we expect 12 (that is, 3x6-6) components for this data set. In analysis, we find that the fifth component's eigenvalue becomes smaller than 1/6, which is .17, making the reliability coefficient α negative. We list, therefore, only the first five components, and adopt the first four components since the contribution of the fifth component drops to a small amount. The information is distributed over the five components as follows: The first four components are all comparatively dominant, accounting for 79 percent of the total information. If we want to analyze the contributions in terms of items as units, the squared item-component correlation is an ideal statistics. Table 7.14 also indicates the percentage of each contribution, with bold face indicating more than 25 percent contributions. We can see from Table 7.14 some patterns of main contributions such as major contributions of blood pressure and migraines and mild contributions of age and anxiety to Component 1, main contributions of

Table 7.14 *Squared Item-Component Correlation*

Component	1	%	2	%	3	%	4	%
Item 1	.92	(**46**)	.38	(19)	.40	(20)	.02	(1)
Item 2	.93	(**47**)	.34	(17)	.48	(24)	.03	(2)
Item 3	.54	(**27**)	.22	(11)	.55	(**28**)	.40	(20)
Item 4	.41	(21)	.29	(15)	.46	(23)	.36	(18)
Item 5	.11	(6)	.52	(**26**)	.18	(9)	.83	(**42**)
Item 6	.36	(18)	.49	(**25**)	.01	(1)	.21	(11)

weight and height to Component 2, contributions of age, migraine and blood pressure to Component 3 and strong contribution of weight and mild contributions of age and anxiety to Component 4.

Another way of looking at item contributions is to examine the inter-item correlation matrices associated with components. Each of these correlation matrices can be subjected to principal component analysis (PCA) to further summarize the inter-item relations. Let us look at correlation matrices associated with the first two components. The correlation matrices of Components 1 and 2 can be decomposed by PCA as in Table 7.19 and Table 7.20, respectively.

The first two PCA components of Component 1 account for 79 percent of the total information, and those of Component 2 for 65 percent. Corresponding coordinates of the six items are listed in Table 7.21, where D1(1) and D1(2) indicate PCA dimensions 1 and 2 of Component 1, and D2(1), D2(2) PCA dimensions 1 and 2 of Component 2. These are projected weights (coordinates). Figures 7.4 and 7.5 are corresponding two-dimensional plots. These plots would help us to determine what kinds of relations have contributed to a particular component. In Figure 7.4, Blood Pressure and Migraine are very close to each other and are the major contributor to the first dimension. By constructing the corresponding contingency table (Table 7.23), we see a strong nonlinear relation between them.

To interpret these component-wise PCA, one helpful way is to construct item-by-item contingency tables in such a way that through rearrangement of rows within rows and that of columns within columns the table shows a linear relation. For example, in Figure 7.4, Age and Anxiety are main contributors to the second dimension of Component 1, and we can create the corresponding contingency table as in Table 7.23, which shows a nonlinear transformation of Age and a linear transformation of Anxiety. Table 7.23 suggests that Age was nonlinearly transformed, while

Table 7.15 *Inter-Item Correlation of Component 1*

1	1.00					
2	.99	1.00				
3	.59	.58	1.00			
4	.47	.51	.67	1.00		
5	.43	.39	.08	-.33	1.00	
6	.56	.57	.13	.19	.20	1.00
	1	2	3	4	5	6

Table 7.16 *Inter-Item Correlation of Component 2*

1	1.00					
2	.06	1.00				
3	.59	-0.31	1.00			
4	.07	.35	.35	1.00		
5	.28	.62	-.01	.19	1.00	
6	.31	.29	.32	.17	.38	1.00
	1	2	3	4	5	6

Table 7.17 *Inter-Item Correlation of Component 3*

1	1.00					
2	.00	1.00				
3	.67	.27	1.00			
4	.26	.54	.17	1.00		
5	-.02	.43	.14	.13	1.00	
6	.22	-.06	.11	.06	-.32	1.00
	1	2	3	4	5	6

Table 7.18 *Inter-Item Correlation of Component 4*

1	1.00					
2	.77	1.00				
3	-.52	-.46	1.00			
4	-.14	-.14	.48	1.00		
5	.17	.22	.50	.32	1.00	
6	.51	.47	-.13	-.11	.42	1.00
	1	2	3	4	5	6

Table 7.19 *Principal Components of Component 1*

Component	1	2	3	4	5	6
Eigenvalue	3.26	1.47	.80	.29	.17	.01
Delta	54.3	24.6	13.4	4.8	2.8	.1
Cum.Delta	54.3	78.9	92.3	97.1	99.9	100.

Table 7.20 *Principal Components of Component 2*

Component	1	2	3	4	5	6
Eigenvalue	2.25	1.67	.96	.65	.35	.13
Delta	37.5	27.7	16.0	10.8	5.9	2.1
Cum.Delta	37.5	65.21	81.21	92.0	97.9	100.

Table 7.21 *Two-Dimensional Coordinates of PCA Components*

Item	D1(1)	D1(2)	D2(1)	D2(2)
1	.96	-.17	.62	.51
2	.96	-.13	.58	-.73
3	.73	.41	.47	.82
4	.64	.71	.54	-.01
5	.33	-.82	.72	-.44
6	.61	-.31	.70	.05

Table 7.22 *Contingency Table of Blood Pressure and Migraines*

Migraine	Rare	Occasional	Frequent
High Blood Pressure	0	0	4
Medium Blood Pressure	3	3	0
Low Blood Pressure	0	0	5

Table 7.23 *Rearranged Contingency Table of Age and Anxiety*

	Low Anxiety	Medium Anxiety	High Anxiety
Old Age	0	0	6
Young	0	1	3
Middle-aged	2	2	1

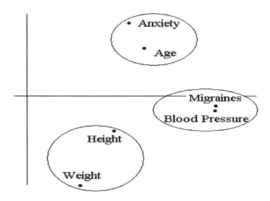

Figure 7.4 *Two Dimensions of Components 1*

Anxiety was not. In Figure 7.5, Blood Pressure and Age appear strong in the second dimension of Component 2, and we have already seen earlier that they are strongly linearly correlated.

Generally speaking, this component-wise PCA tends to provide strong one-sided (the sign of weights) first dimension as we see that all the data points in Figures 7.4 and 7.5 appear in the first and the fourth quadrants. Since each inter-item correlation matrix represents a single component, the above observation may be expected. If so, component-wise PCA may be of help only with identifying what linear or nonlinear transformations of individual items may have contributed to creating a component.

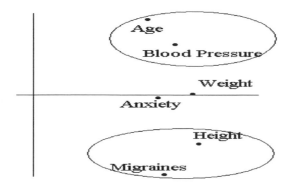

Figure 7.5 *Two Dimensions of Components 1*

7.6 Further Discussion

7.6.1 Evaluation of *alpha*

Using the same example of Blood Pressure, let us look at the reliability coefficient α. The six items are all answered in terms of three ordered categories, and this kind of data is often analyzed by assigning integers such as 1, 2, 3 to, for example, low anxiety, medium anxiety and high anxiety, respectively. This scoring scheme is referred to as Likert score, or Likert-type scoring (Likert, 1932). To find out how effectively MUNDA extracts information from data, let us consider (1) the value of *alpha* we obtain from Component 1 of MUNDA, (2) the value from Likert-type scoring and (3) the value obtained from the first component of PCA. These procedures are based on:

(1) Option weights determined so as to maximize α.

(2) Option weights given by Likert-type scoring.

(3) Optimal linear combination of Likert-type item scores.

The corresponding values of α are listed in Table 7.24. We also obtain the effective test length, h, and the equivalent number of questions, $n(E)$: In Table 7.24, we used MUNDA as the criterion. The results show that with Likert-type scoring of options we need 46 "parallel" (roughly speaking, mathematically comparable) items to reach the value of α attained by MUNDA, and that with PCA of Likert-type scoring we need 18 items.

Table 7.24 *Effective Test Length*

Method	Quantification	Likert-Type	PCA
α	.83	.39	0.61
h	1.0	7.7	3.1
Equivalence: e(E)	6 items	46 items	18 items

In the principal coordinate system, each variable in PCA is represented as a straight line (axis), as demonstrated earlier, while categories of each variable in our method no longer lie on a straight line. In the above comparison, however, we used only the first component, meaning that all weights for options were lined up on a unidimensional continuum. Even then, the analysis reveals that MUNDA taps into much more information than the other two procedures.

As a brain wave analyzer filters a particular wave such as alpha, most statistical procedures, in particular PCA, factor analysis, multidimensional scaling and other correlational methods, play the role of a linear filter, meaning that they filter out most of the information in the data which we call a nonlinear portion of the data. In this context, MUNDA should be re-evaluated and highlighted as a valuable means for analyzing both linear and nonlinear information in the data, in particular, in the behavioral sciences where it seems that nonlinear relations are more abundant than linear relations.

7.6.2 Standardized Quantification

Considering that the option frequencies and the number of options have great impacts on the amount of information in data, Nishisato (1991, 1994) considered what he calls "standardized dual scaling." Two schemes of standardization were proposed:

(A) Option standardization: In this case, the contributions of all options of all items are set equal in the total space. This can be done by dividing the columns of matrix **B**, discussed earlier, by the square root of SS(jp), that is,

$$\sqrt{SS(jp)} = \sqrt{n(N - f_{jp})} \qquad (7.26)$$

This standardization is used if MUNDA is formulated in terms of frequencies as in dual scaling. If it is formulated in terms of proportion as in correspondence analysis, the above expression should be changed to

$$\sqrt{SS(jp)} = \sqrt{\frac{N - f_{jp}}{nN}} \qquad (7.27)$$

(B) Item standardization: The contributions of items are equated, no matter how many options they may have, by dividing the columns of **B** by the square root of SS(j), that is, if the formulation of MUNDA is in terms of frequencies,

$$\sqrt{SS(j)} = \sqrt{nN(m_j - 1)} \qquad (7.28)$$

or if it is in proportion,

$$\sqrt{SS(j)} = \sqrt{\frac{m_j - 1}{n}} \qquad (7.29)$$

He analyzed Tung's full data (see Nishisato (1994, 1996)). Table 7.25 shows four sets of scores of the first fifteen subjects, corresponding to Component 1, Component 2 of MUNDA (indicated by C1 and C2, respectively), Component 1 of option standardization (OS) and Component 1 of item standardization (IS).

The findings are remarkable. Once we standardize the analysis, be it for options or items, the distribution of subjects' scores changes dramatically, and it no longer resembles that of Component 1 of the original data. The first components of option-standardization and item-standardization, however, look very much like Component 2 of the original data. Nishisato (1991) tentatively concluded that standardization has the effect of suppressing the effects of outlier responses, that those three responses of Subject 20 are responsible for creating Component 1, the same finding as we saw earlier in this chapter, using a subset of the data. He also noted that unfortunately, however, standardization destroys optimal properties of the quantification procedure (e.g., maximization of α). This leaves a number of questions unanswered and suggests that the quantification of standardized data should be further investigated.

The standardization problem is indeed very important, for even when we can control the number of options of all items by the design phase of the data collection instrument it is not possible to control the frequencies of options of items. When all the options are ordered, one can merge an option with a very small frequency with the adjacent option, which would partly solve the problem. However, when options are nominal, there is no rational way to merge two options. At this point, option standardization becomes important. One should also note that

Table 7.25 *Effects of Standardization*

Subj	C1	C2	OS	IS
1	.69	-.82	-1.32	-1.03
2	.36	-1.20	-1.40	-.92
3	.45	-1.00	-1.32	-.92
4	1.76	.12	-.98	-1.37
5	.41	-.91	-1.28	-.98
6	.38	-.16	-.47	-.36
7	.56	-.93	-1.36	-.89
8	.41	-.68	-.94	-.53
9	.49	-.69	-1.06	-.75
10	.52	-.59	-.98	-.75
11	-.85	1.17	1.15	1.53
12	-.09	-.01	-.17	.31
13	-.89	.63	.81	1.40
14	-1.12	1.71	1.62	1.79
15	-.96	1.53	1.66	1.87

standardization might also bring to us additional advantages which may overweigh the disadvantage of losing optimality, that is, bringing us ease with which some mathematics is attained, for example, properties of metric and geometric regularities. At the current moment, we must leave this standardization matter for the future as an important and promising venue of investigation.

CHAPTER 8

Sorting Data

8.1 Example

Sorting data consist of a table of *n* objects by *N* respondents of chosen classification piles. The following is an example in which 4 countries (a-d) were sorted into 2, 3, 3, 4 and 2 piles of similar countries by five subjects (S1-S5), under the instruction that "Sort the following 7 countries into as many piles as you see appropriate, using your judgment that countries in each pile are similar among themselves and different from countries in other piles." The subjects were asked to put 1 for the first country presented, and then go down the list, assigning 1 to any number of countries which the subject considers are similar to the first country; assign 2 to one country from the remaining countries, and assign 2 to any number of countries not already chosen which the subject thinks are similar to the country 2; and so on until all the countries are classified into piles.

Table 8.1 *Sorting Data*

Country	S1	S2	S3	S4	S5
a	1	1	1	1	1
b	2	2	3	3	2
c	1	1	1	2	1
d	1	3	2	4	1

8.2 Early Work

Sorting data are known to capture a lot of information from subjects about their judged objects (Nagy, 1984). However, the fact that each subject can use as many piles as one likes and the fact that each pile size can vary from pile to pile and from subject to subject present tricky problems for data analysis, and it becomes very difficult to identify an appropriate method for analyzing sorting data.

It seems that Wiley's latent partition analysis (Wiley, 1967) is the first method ever proposed to handle sorting data. His model is based on the idea of latent classes for classification and the task is to assign probabilities for each variable to belong to each latent class. Although it is an attractive framework, it is understood that numerical methods need to be improved, considering that some estimates of probability measures may go beyond the boundaries of 0 and 1.

Evans (1970) reformulated, extended and generalized Wiley's procedure. Evans' approach, too, is probabilistic rather than descriptive. His several models offer interesting ways to conceptualize sorting behavior, but are definitely different from MUNDA's approach. For details, interested readers are referred to Wiley (1967) and Evans (1970).

Years later, Takane (1980b, 1984) demonstrated that sorting data could be analyzed by dual scaling simply by transposing the data, or exchanging the roles of subjects and item options in multiple-choice data with objects and subjects' piles in sorting data, respectively. Please study the above example in which the rows of the table consist of countries (rather than subjects as in multiple-choice data) and the columns are for subjects (rather than items as in multiple-choice data). The number of piles used by a subject in sorting data is treated as the number of options in multiple-choice data. With these changes for analysis, the rest is a straightforward application of MUNDA used for multiple-choice data. Therefore, we will follow Takane's procedure.

There are a few points to note. T(comp) and T(inf) can be defined accordingly: N is the number of objects, m_j the number of piles Subject j used for sorting, and n the number of subjects. In sorting data, we obtain an inter-subject correlation matrix for each component, which tells us similarities and differences among subjects. We can plot objects (or stimuli), but not subjects. Recall the reversed situation for multiple-choice data: we can plot subjects, but not items. Instead of item statistics, we can now look at subjects (judges) statistics. For instance, we can find how individual judges are contributing to each component in terms of the sums of squares of the subject-component correlation coefficients. All of these will become clear as we look at an example.

8.3 Sorting Familiar Animals into Clusters

Data were collected at a university in Nishinomiya, Japan, when the author was teaching there. Thirty-five animals familiar to the subjects

Table 8.2 *Animals to be Sorted into Clusters*

Dog (A0)	Alligator (A1)	Chimpanzee (A2)	Cow (A3)
Crow (A4)	Pigeon (A5)	Cheetah (A6)	Chicken (A7)
Bear (A8)	Cat (A9)	Rabbit (B0)	Frog (B1)
Goat (B2)	Tiger (B3)	Rhinoceros (B4)	Giraffe (B5)
Duck (B6)	Sparrow (B7)	Hippopotamus (B8)	Monkey (B9)
Turkey (C0)	Pig (C1)	Crane C2)	Leopard (C3)
Ostrich (C4)	Lizard (C5)	Horse (C6)	Racoon (C7)
Tortoise (C8)	Snake (C9)	Lion (D0)	Elephant (D1)
Camel (D2)	Hawk (D3)	Fox (D4)	

(Table 8.2) were sorted into piles of similar animals by 15 subjects (Table 8.3). The subjects used the following numbers of piles: From Subject 1 to 15, the numbers are 8,3,9,10,6,5,7,5,5,5,6,4,8,8 and 6, respectively. We should note that the number of piles is free for each subject to decide, but that this freedom can create a problem of uneven contributions by different subjects to the determination of clusters. In multiple-choice data, we noted that the contribution of each item can be measured by the sum of squared item-component correlation over the entire space, which was equal to the number of options of the item minus 1. Thus, the number of piles minus 1 is the contribution of each subject to the outcome of analysis. If the distribution of the number of piles is uneven, therefore, it is possible that one subject with a large number of piles may dominate in deciding the characteristics of many components.

The current data contain 80 components (i.e., the total number of piles minus the number of subjects = 95-15=80). Thus, the contributions of the 15 subjects to the 80 components are as in Table 8.4, in which we enter the contributions of only the first 5 components. It is interesting to note that substantial individual differences show up from component 3. Note also the point mentioned earlier: Subject 2 contributes to the total information by 2, while Subject 4 by 9. This difference shows up from component 3: note the contributions of Subject 2 to components 3, 4, 5, which are all zero! This warns us that if we want all the subjects to contribute to the information evenly, we must impose on their sorting task that the number of piles should be fixed at a reasonable constant.

Let us now look at the contributions of the first ten components (Table 8.5). Since there are 80 components, we can tell from Table 8.5 that the information is very thinly distributed over the remaining 70 components.

Table 8.3 *Classification of Animals*

Ss	1	2	3	4	5	6	7	8	9	10	11	12	13	14	15
A0	1	1	1	1	1	1	1	1	1	1	1	1	1	1	1
A1	2	2	2	2	2	2	3	3	2	2	2	2	3	2	8
A2	3	1	3	3	3	1	4	3	3	3	3	2	3	2	3
A3	4	1	4	4	4	2	3	4	1	2	1	1	1	3	4
A4	5	3	5	5	5	4	2	5	4	4	4	4	4	6	5
A5	5	3	5	5	5	4	2	5	4	4	4	1	4	6	5
A6	6	1	1	6	1	2	3	4	5	1	3	1	2	4	2
A7	5	3	5	10	5	4	6	5	4	4	1	4	4	6	5
A8	1	1	6	4	2	2	3	3	5	5	5	1	1	4	2
A9	1	1	1	6	1	2	1	1	1	1	1	1	2	1	1
B0	1	1	7	4	1	5	7	1	1	1	1	1	5	1	1
B1	2	2	8	9	6	3	5	2	2	2	6	3	6	7	6
B2	6	1	4	4	4	5	7	4	1	1	1	1	5	1	1
B3	3	1	1	6	1	2	3	4	5	1	2	1	2	4	2
B4	7	1	3	4	2	5	3	4	5	2	5	1	7	5	2
B5	4	1	9	7	4	5	3	4	5	3	5	1	7	5	4
B6	5	3	5	10	5	4	2	5	4	4	4	4	4	6	5
B7	5	3	5	5	5	4	2	5	4	3	4	4	4	6	5
B8	7	1	2	4	2	5	3	2	5	2	5	1	7	5	2
B9	1	1	3	3	3	1	4	3	3	3	3	2	3	2	3
C0	5	3	5	10	5	4	6	5	4	4	1	4	4	6	5
C1	4	1	4	4	4	2	7	4	1	1	1	1	5	3	1
C2	5	3	5	5	5	4	2	5	4	4	4	4	4	6	5
C3	4	1	1	6	1	2	3	4	5	1	2	1	2	4	2
C4	6	3	5	10	5	4	3	5	4	4	5	4	4	5	4
C5	8	2	8	9	6	3	5	2	2	2	6	3	8	7	6
C6	1	1	4	4	4	1	3	4	1	1	1	1	5	3	4
C7	1	1	1	1	0	2	7	1	1	1	2	1	1	1	1
C8	2	2	8	2	6	3	5	2	2	5	6	3	6	7	6
C9	2	2	8	2	2	3	5	2	2	2	6	3	8	7	6
D0	3	1	1	6	2	2	3	4	5	1	2	1	7	4	2
D1	7	1	2	7	4	5	3	4	5	5	2	1	7	5	2
D2	7	1	4	7	4	5	3	4	5	3	5	1	7	5	4
D3	5	3	5	5	5	4	2	5	4	4	2	4	4	6	5
D4	1	1	1	6	1	2	7	1	1	1	2	1	1	1	1

Let us now look at clusters of animals. Since the weights for piles of subjects are of no special interest, we will look at plots of only animals, Components 1 versus 2, 1 vs. 3 and 1 vs. 4. The reason why we want to plot Components 2, 3 and 4 against Component 1 is for us to see how Components 2, 3 and 4 contribute individually. Namely, once Component 1 is fixed, we may pay attention only to what additional move (i.e., ups and downs) each of the other components will bring to the

Table 8.4 *Contributions of Subjects on Clustering Animals*

Subject	1	2	3	4	5	6	7	8
Comp.1	.95	.98	.99	.98	.96	.98	.93	.97
2	.90	.90	.98	.93	.81	.95	.95	.82
3	.15	.00	.67	.96	.95	.41	.96	.62
4	.68	.00	.53	.59	.42	.34	.70	.52
5	.21	.00	.79	.64	.57	.35	.14	.03
.	.	.	.	.	.	.	.	.
.	.	.	.	.	.	.	.	.
80	.	.	.	.	.	.	.	
Sum	7.	2.	8.	9.	5.	4.	6.	4.

Subject	9	10	11	12	13	14	15	Sum
Comp.1	.98	.83	.67	.88	.99	.95	.95	14.01
2	.91	.48	.83	.91	.97	.98	.95	13.27
3	.97	.41	.62	.79	.80	.99	.96	10.28
4	.63	.34	.53	.01	.67	.91	.68	7.56
5	.26	.08	.45	.01	.61	.79	.58	5.51
.		.	.	.	.	.		
.		.	.	.	.	.		
80		.	.	.	.	.	.	
Sum	4.	4.	5.	3.	7.	7.	5.	80.00

sorting patterns. Let us start with the joint plot of Component 1 against Component 2 (Figure 8.1).

As we can see from Table 8.4, most subjects have high contributions to the first two components, meaning that they have agreed with clustering certain animals. From Figure 8.1, those clusters are:

- Crow, sparrow, pigeon, duck, chicken, crane, hawk
- Lizard, frog, tortoise, snake
- Ostrich
- Alligator
- Hippopotamus

As for the remaining 20 animals, the subjects are not consistent in their

Table 8.5 *Distribution of Information over Components*

Component	1	2	3	4	5
Eigenvalue	.93	.88	.69	.50	.37
Singular Value	.97	.94	.83	.71	.61
Delta	17.7	16.8	13.0	9.6	7.0
Cum. delta	17.7	34.4	47.5	57.0	64.0

Component	6	7	8	9	10
Eigenvalue	.25	.23	.17	.15	.15
Singular Value	.50	.48	.41	.39	.39
Delta	4.7	4.3	3.2	2.9	2.8
Cum. delta	68.6	72.9	76.1	79.0	81.8

sorting and those animals were put together without clear-cut features. When we plot Component 3 against Component 1 (Figure 8.2), we can see that the contribution of Component 3 is to isolate chimpanzee and monkey out of those 20 animals.

- Chimpanzee, monkey

We then move on to plotting Component 4 against Component 1 (Figure 8.3), and we discover that Component 4 contains information about further decomposing of the mixed group of animals into clusters.

- Dog, cat, racoon, rabbit, fox
- Rhinoceros, hippopotamus, giraffe, elephant, camel
- Lion, bear

With these four components, 57 percent of information is explained. We can go on to look into further clustering of animals. One practical problem with sorting data, however, is that the information is so widely and thinly distributed over many components that by the time we explain 80 percent of information we must look at many more components than the typical case of quantification. Imagine that 80 components are involved in sorting of only 34 animals!

Another astonishing aspect peculiar to sorting data is that the inter-subject correlation associated with the first few components tends to be very high, which we do not expect to see in multiple-choice data. Remember that in multiple-choice data, the correlation is between items, while in

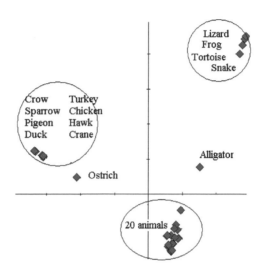

Figure 8.1 *Clusters of Animals by Components 1 and 2*

sorting data it is between subjects (judges). To see this aspect, let us look
at the correlation between subjects associated with the first component
(Table 8.6).

The very high average correlation between subjects is worth noting.
This phenomenon is partly due to the fact that all subjects are judging
the same set of animals with respect to the degrees of their similarities,
which is quite different from judging subjects' own personality traits or
taste preferences. A possibly more important reason for high correlation
than the above is because the subjects use many more piles in sorting
data than a typical number of options of a multiple-choice item. In the
current data set, one subject, for example, used 10 piles. It would be
rather unusual to see a multiple-choice item with 10 options. Remember
that MUNDA is a descriptive procedure that takes advantage of available

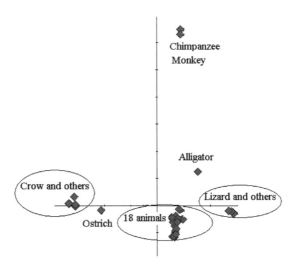

Figure 8.2 *Clusters of Animals by Components 1 and 3*

freedom for optimization. So, the more piles the data have the greater
the correlation between subjects is expected. Recall that due to the large
number of piles subjects used, the current data yielded 80 components!

8.4 Some Notes

Sorting data we have seen indeed reflect freedom of judgment as is well
known among some psychologists. From a data analytic point of view,
however, this freedom seems to act as hindrance to our quest for
understanding data. As is clear from the current example, one may have
to sacrifice the freedom associated with the sorting task, and the
investigator may have to impose a condition on the sorting task, in
particular, the number of piles that one is expected to use. Or, researchers
may have to indicate the upper limit of the total number of piles one may
use, for it is always the case that when subjects see some unfamiliar

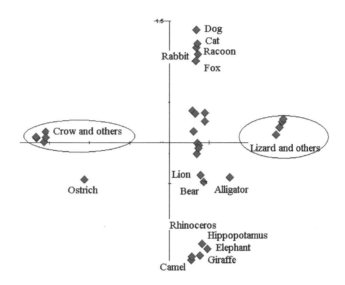

Figure 8.3 *Clusters of Animals by Components 1 and 4*

objects (animals) they tend to create piles with single entries, that is, one pile for one animal, for example. This must be avoided since we are looking for links among objects, not to mention an enormous number of components one may have to look at.

Mayenga (1997) investigated free sorting and restricted sorting, the latter being sorting under a fixed number of piles. Interested readers are referred to his thesis, but his conclusion is that the final clusterings under different conditions are very similar, and therefore that the restricted sorting task, either setting the constant number of piles one may use or setting the range in the number of piles, may be a good alternative so as to avoid dealing with an unmanageable number of clusters or components. Mayenga (1997) demonstrated that even eight theoretical clusters of animal classification by the Linnaean system could be identified from the restricted sorting of six piles per subject. This finding was, of course, based on the situation with substantial enough

Table 8.6 *Between-Subject Correlation of Component 1*

	1	2	3	4	5	6	7	8	9	10	11
1	100										
2	96	100									
3	95	99	100								
4	96	100	99	100							
5	94	98	99	98	100						
6	95	99	100	99	98	100					
7	97	93	94	94	93	94	100				
8	94	98	99	98	98	98	93	100			
9	96	100	99	100	98	99	93	98	100		
10	87	92	92	91	92	91	84	92	92	100	
11	81	79	79	79	77	79	84	77	77	63	100
12	89	94	94	93	93	94	88	93	94	83	72
13	95	100	100	99	98	100	94	99	99	91	80
14	98	95	95	95	93	95	99	93	95	85	83
15	97	94	96	95	94	96	99	94	94	86	84

	12	13	14	15
12	100			
13	95	100		
14	89	95	100	
15	89	95	99	100

individual differences to interlink animals sorted into different piles. Individual differences are indeed crucial to create eight clusters from sorting of animals into a smaller number of piles than eight.

CHAPTER 9

Forced Classification of Incidence Data

In Chapter 8, we discussed the details of quantification of multiple-choice data. The object was to determine category (option) weights so as to maximize the correlation ratio or Cronbach's internal consistency reliability coefficient α. This optimization was carried out in terms of the entire set of items. In this context, the category weights of one item are likely to change if another item was omitted or a few other items are added to the data set. This may not always be what we want. Consider the following questions:

1. Can we not carry out quantification in such a way that all the characteristics of one item are uniquely determined, independently of what other items are included in the data set?

2. Can we not maximize the contribution of a particular item to the first component, rather than that of all the items?

3. Can we not maximize the correlation of items with a particular item in the data set, rather than the average of all the correlation coefficients?

All of these questions can be answered with 'yes' if we use forced classification.

Nishisato (1984a) proposed a simple way to carry out discriminant analysis with categorical data, which he called forced classification for the reason to become clear later.

9.1 Early Work

Since quantification is based on singular value decomposition of the data, the axis orientation is determined by the structure of the entire data. There are occasions in which the investigator wishes to determine the axis orientation in terms of a particular variable. If so, this task is handled by forced classification, or more popularly called discriminant analysis. Nishisato's procedure is based on multiplying a variable of choice (the criterion) by a large enough constant, called forcing agent, to make it dominant in quantification. Its implications for data analysis are often over shadowed by more popular eigenvalue decomposition of data structure, and remain overlooked partly because of its simplicity and partly because we are so used to procedures based on eigenvalue decomposition. The original procedure was generalized (Nishisato, 1986a), further clarified (Nishisato, 1988a) and was considered for market segmentation (Nishisato, 1988b; Nishisato and Gaul, 1990). In an overview of dual scaling (Nishisato, 1996), he characterized forced classification as 'a gold mine of dual scaling.' Nishisato and Baba (1999) provided an additional contribution to the development of the procedure.

9.2 Some Basics

Let us denote the data of n multiple-choice questions from N subjects as

$$\mathbf{F} = [\mathbf{F}_1, \mathbf{F}_2 \cdots \mathbf{F}_j \cdots \mathbf{F}_n] \qquad (9.1)$$

where $\mathbf{F}_j$ is an N-by-m_j matrix, the row i consisting of subject i's response to item j, 1 being the choice and 0 the non-choice out of m_j options. Typically $\mathbf{F}_j \mathbf{1} = \mathbf{1}$, meaning that each subject chooses only one option per item. The procedure of forced classification is based on two principles, principle of equivalent partitioning, PEP, and principle of internal consistency, PIC.

9.2.1 Principles PEP and PIC

Let us introduce two key principles used in forced classification. The principle of equivalent partitioning (Nishisato, 1984a) is demonstrated in the following numerical examples of equivalent data structures (Tables 9.1 to 9.3). Notice that the structure of the data in Table 9.1 remains the

Table 9.1 *Original 3x3 Table with Optimal Weights*

	1	2	3	Row Weight
1	30	10	10	1.78
2	15	30	45	-.47
3	10	40	20	-.67
Column Weight	1.67	-.67	-.51	$\eta^2 = .1859$

same, that is, the same eigenvalue and the same set of distinct weights, even when a row is partitioned into proportional rows and a column into

Table 9.2 *Partitioned Tables*

30	10	10	1.78	15	5	5	1.78
5	10	15	-.47	15	5	5	1.78
10	20	30	-.47	15	30	45	-.47
10	40	20	-.67	5	20	10	-.67
				5	20	10	-.67
1.67	-.67	-.51	$\eta^2=.1859$	1.67	-.67	-.51	$\eta^2=.1859$

Table 9.3 *Partitioned Table*

30	5	5	10	1.78
15	15	15	45	-.47
5	10	10	10	-.67
5	10	10	10	-.67
1.67	-.67	-.67	-.51	$\eta^2=.1859$

proportional columns (Tables 9.2 and 9.3). Thus, if we see, for example, three proportional rows in the data matrix, we can combine those rows into a single row without changing the data structure. The above property is referred to as the *principle of equivalent partitioning* (PEP), and this property of quantification is also well known in correspondence analysis under the name of the principle of distributional equivalence (e.g., Benzécri et al., 1973). The same principle applies to the distance in correspondence analysis (Escofier, 1978).

Let us now introduce the *principle of internal consistency* (PIC). Suppose that we repeat $\mathbf{F}_j$ k times in the data matrix, which we indicate by $f(k\mathbf{F}_j)$,

$$f(k\mathbf{F}_j) = [\mathbf{F}_1, \mathbf{F}_2 \cdots \overbrace{\mathbf{F}_j, \mathbf{F}_j \cdots \mathbf{F}_j}^{\text{repeated } k \text{ times}}, \cdots \mathbf{F}_n] \qquad (9.2)$$

Let us also introduce another matrix, $g(k\mathbf{F}_j)$, in which $\mathbf{F}_j$ is not repeated k times, but in which all the elements of 1s in $\mathbf{F_j}$ are replaced with ks, that is, $k\mathbf{F}_j$,

$$g(k\mathbf{F}_j) = [\mathbf{F}_1, \mathbf{F}_2, \cdots, k\mathbf{F_j}, \cdots, \mathbf{F_n}] \qquad (9.3)$$

The principle of equivalent partitioning assures us that the structures of the two matrices $f(k\mathbf{F_j})$ and $g(k\mathbf{F_j})$ are identical. In other words, the distinct optimal weight vectors for the rows of one matrix are exactly the same as those of the other matrix, and the eigenvalues of the two matrices are also the same. For this property, it would be easier to deal with the latter matrix $g(k\mathbf{F_j})$ than the former $f(k\mathbf{F_j})$ because the dimension of the matrix in $g(k\mathbf{F_j})$ is the same as that of the original matrix $\mathbf{F}$.

As k increases in the above expression, the response patterns in $\mathbf{F}_j$ become more dominant in the data set, and eventually we will see that the response patterns in the repeated $\mathbf{F}_j$ determine the first component. Notice that the computation involved here is ordinary MUNDA with an altered data matrix by multiplying the chosen submatrix by a large enough scalar k. Consider increasing k to infinity. The asymptotic properties of this multiplication process are

$$\lim_{k\to\infty} r_{jt}^2 = 1 \tag{9.4}$$

$$\lim_{k\to\infty} \rho^2 = 1 \tag{9.5}$$

For Item j with m_j options, the first $(m_j - 1)$ components attain the above asymptotic results. These components are called *proper components* of forced classification, and item j is referred to as the *criterion item*.

Thus, forced classification is defined as a procedure of quantifying $g(k\mathbf{F}_j)$ with a constant k large enough to yield those asymptotic results for the first $(m_j - 1)$ components.

To clarify the above description, let us look at a numerical example from Nishisato and Baba (1999). They used an artificial set of data as shown in Table 9.4 in the response-pattern format, which appeared in Nishisato and Baba (1999, Table 1, p.209).

They compared the results from dual scaling (MUNDA) and forced classification in terms of the squared item-total (component) correlation. The numbers of options of Items 1, 2, 3, 4, and 5 are 2, 3, 4, 5, 6, respectively, and the forced classification procedure was carried out with Item 4 as the criterion. Since Item 4 has five options, there are four proper components.

Let us start with ordinary quantification (Table 9.5) and then compare the results with those from forced classification with Item 4 as the criterion and conditional analysis with the forcing agent k of 10,000 (Table 9.6).

First, notice that in both tables the sum of the squared item-total correlation over 15 components of each item is equal to the number

Table 9.4 *Five-Item Multiple-Choice Data*

Ss	1	2	3	4	5
1	10	100	1000	10000	100000
2	01	100	0100	00100	000010
3	10	010	0010	00010	000001
4	01	001	0001	00001	000001
5	10	010	0010	01000	010000
6	01	100	0001	00100	000100
7	10	001	0001	10000	000100
8	01	010	1000	10000	100000
9	10	001	0100	00001	010000
10	01	100	1000	01000	000010
11	01	010	0100	01000	010000
12	10	010	0010	00010	000010
13	01	001	0001	00001	000001
14	10	001	0001	00100	001000
15	10	100	1000	10000	000100
16	01	001	0100	01000	000100
17	10	010	1000	00001	010000
18	01	100	0010	00010	001000

of options minus 1. This sum is a measure of the item contribution to the data set. Secondly, notice the distribution of the squared item-total correlations of Item 4 in the two tables. In Table 9.5 of ordinary MUNDA, the information of Item 4 is distributed over 15 components. In Table 9.6 of forced classification, the criterion item (Item 4) is exhaustively explained by the first four proper components. Thus, if we are to seek relations between the criterion item and each of the remaining items, the relevant statistics under forced classification are all included in the first four rows of Table 9.6. What are the remaining components where the criterion item has no contributions?

9.2.2 Conditional Analysis

Data typically contain more information than accounted for by the proper components. The extra components generated by forced classification

Table 9.5 *Squared Item-Total Correlation (Standard Dual Scaling)*

Component	Item 1	Item 2	Item 3	Item 4	Item 5
1	.072	.726	.868	.497	.480
2	.040	.445	.582	.746	.681
3	.187	.007	.631	.685	.631
4	.081	.252	.231	.794	.690
5	.303	.033	.050	.142	.809
6	.029	.009	.043	.306	.575
7	.158	.059	.036	.143	.456
8	.020	.206	.065	.245	.136
9	.002	.013	.325	.184	.090
10	.070	.131	.016	.089	.226
11	.032	.081	.025	.050	.096
12	.001	.033	.023	.069	.075
13	.003	.003	.077	.030	.051
14	.002	.002	.027	.021	.003
15	.000	.001	.001	.002	.002
Sum	1.0	2.0	3.0	4.0	5.0

are called *conditional components*, for these are obtained after we eliminate the effects of the criterion variable. In other words, those statistics below the fourth row of Table 9.6 are free from the influence of the criterion item, as shown by zero values of the criterion-item total correlation.

The conditional analysis of forced classification by Nishisato (1984a) is also referred to as partial correspondence analysis (Yanai, 1986, 1987, 1988; Maeda, 1996, 1997; Yanai and Maeda, 2002) and partial canonical correspondence analysis (Ter Braak, 1988). It is also closely related to quantification with external variables (Takane and Shibayama, 1991) and dual scaling of multiway data matrices (Nishisato, 1971, 1972, 1980a, 1994; Poon, 1977; Lawrence, 1985; Nishisato and Lawrence, 1989).

In the ordinary MUNDA, we maximize the item-total (item-component) correlation, which we can see from the relation that the eigenvalue is equal to the average of the squared item-component correlation. In forced classification, the criterion item becomes the component itself, meaning that option weights of non-criterion items are

Table 9.6 *Squared Item-Total Correlation (Forced Classification)*

Component	Item 1	Item 2	Item 3	Item 4	Item 5
1	.005	.126	.701	1.000	.321
2	.065	.198	.249	1.000	.520
3	.066	.100	.129	1.000	.317
4	.013	.049	.197	1.000	.177
5	.017	.550	.577	.000	.747
6	.508	.370	.086	.000	.616
7	.027	.147	.291	.000	.543
8	.001	.080	.011	.000	.798
9	.115	.125	.472	.000	.127
10	.100	.051	.019	.000	.454
11	.073	.079	.030	.000	.158
12	.006	.116	.066	.000	.131
13	.001	.003	.108	.000	.077
14	.003	.005	.065	.000	.015
15	.001	.003	.001	.000	.003
Sum	1.0	2.0	3.0	4.0	5.0

determined so as to maximize their correlations with the criterion item. Conditional components reflect the other extreme, namely those which are independent of the criterion item.

9.2.3 Alternative Formulations

Nishisato (1984a) states that proper components of forced classification can also be obtained in the following two ways:

1. MUNDA of the contingency table constructed as the options of the criterion items by the options of the remaining items, that is,

$$\mathbf{F}_j'[\mathbf{F}_1, \mathbf{F}_2, \cdots, \mathbf{F}_{j-1}, \mathbf{F}_{j+1}, \cdots, \mathbf{F}_n] = \mathbf{C}_{j,t-j} \qquad (9.6)$$

2. MUNDA of the data matrix projected onto the space spanned by the columns of the criterion item, that is,

$$\mathbf{F}_j(\mathbf{F}_j'\mathbf{F}_j)^{-1}\mathbf{F}_j'\mathbf{F} = \mathbf{P}_j\mathbf{F} \qquad (9.7)$$

Table 9.7 *The Criterion-by-Remaining Item Contingency Table*

3	1	2	1	1	3	0	0	1	2	0	1	1	0	0
1	3	1	2	1	1	2	1	0	0	2	0	1	1	0
1	2	2	0	1	0	1	0	2	0	0	1	1	1	0
2	1	1	2	0	0	0	3	0	0	0	1	0	1	1
2	2	0	1	3	1	1	0	2	0	2	0	0	0	2

where the projection operator P_j is given by

$$P_j = F_j(F_j'F_j)^{-1}F_j' \qquad (9.8)$$

In these two cases, no asymptotic results are obtained, but the results are always exact.

As for extracting conditional components, it is not immediately clear how one can obtain them from the contingency format of 1. From the projection method of 2, however, they are obtainable from MUNDA of the matrix $(I - P_j)F$, where I is the identity matrix.

9.2.4 Adjusted Correlation Ratio

Up to this point, we have discussed forced classification as quantification of an *altered* data matrix. Thus forced classification does not yield by itself the statistic that corresponds to eigenvalues or the correlation ratios as we recall that all the asymptotic values of these statistics are 1.

The question then is how to assess the contribution of each proper component to the total information. Recall that there are $(m_j - 1)$ proper components and that each of them yields asymptotically ρ^2 of 1. This statistic, however, is an outcome of forcing the process with a large constant k to a limit, and it does not mean that each proper component has the same amount of contribution to the total information. What is then the contribution of each proper component? To respond to this query, let us use the property that forced classification is equivalent to quantification of the contingency table of the options of the criterion variable by the options of the remaining items, which is in our example (Nishisato and Baba, 1999) the contingency table as shown in Table 9.7.

The squared item-component correlations of the five components of the contingency table are as shown in Table 9.8. Using this table as an example, Nishisato and Baba (1999) provided the following formula for

Table 9.8 *Squared Item-Component Correlation*

Item	Comp1	Comp2	Comp3	Comp4
1	.004	.066	.065	.013
2	.127	.195	.101	.049
3	.705	.247	.128	.196
4	1.000	1.000	1.000	1.000
5	.315	.524	.318	.178
*	.288	.258	.126	.109

the correlation ratio in the subspace of the criterion variable,

$$\rho_k^2 = \frac{\sum_{j=1}^{n} r_{jt(k)}^2 - 1}{n - 1} \tag{9.9}$$

In Table 9.8, the entries in the last row, marked *, are the values calculated by this formula.

The average of the squared item component correlation of the non-criterion variables in Table 9.6 should be equal to those adjusted values in Table 9.8, but for some reason or other we see some discrepancies. These may be due to the approximation with k, the forcing agent, or to some computational errors.

Once the adjusted correlation ratios are calculated, the statistic δ for each of the forced classification proper components must be re-defined accordingly, that is, in terms of these adjusted statistics.

9.2.5 *Value of Forcing Agent*

Many researchers have asked the question: "What is an appropriate value of k?" Nishisato (1986) showed a graph in which the average correlation of the criterion item with the remaining items increases quickly as a function of k, reaching an asymptotic value even when the value of k is as small as 5. Since the value of k is equivalent to the number of repetitions of the criterion variable in the date set, however, the question is when the criterion variable becomes 'dominant' in the given data set. Generally speaking, the value of k must be greater as the number of items increases. There is another twist with the question for an appropriate value of k: 'dominance' needs to be interpreted differently for different statistics. If the correlation ratio, which is the average of the squared item-component correlation coefficients, is used to judge the dominance, it is comparatively insensitive to minute discrepancies in option weights, hence a comparatively small value of k may lead to the squared item-total correlation close to one. In contrast, if the investigator looks at discrepancies in corresponding weights from a particular value of k, a much larger value of k is needed for the dominance judgment than the case with such a summary statistic as correlation ratio. To see negligible differences in weights, Nishisato and Baba (1999) showed that even the value of 10,000 may not be sufficient for their sample data as shown in Table 9.4. They also showed with a numerical example that, in forced classification, option weights of non-criterion items are projected weights and options of the criterion item normed weights. In other words, forced classification is a procedure to project options of non-criterion items onto the space spanned by the options of the criterion item. This, of course, is obvious from the fact that forced classification is MUNDA of P_jF.

9.3 Age Effects on Blood Pressures and Migraines

Since we have looked at the data on blood pressure, migraines, age and some other variables in Chapter 1, let us use the same data (Table 1.10) and carry out forced classification analysis with the variable 'age' as the criterion. Since the criterion item has three options (three age groups), there will be two proper forced classification components. Since the program DUAL3 (Nishisato and Nishisato, 1994) stops extracting components once the value of reliability α becomes negative, we can not look at all 12 possible components, that is, 18 (options)-6 (items), but only three components, two of which are proper and the other one conditional (Tables 9.9 and 9.10).

Both eigenvalues and singular values of the first two proper solutions

Table 9.9 *Some Key Statistics*

Component	1	2	3
Eigenvalue	.994	.993	.004
Singular Value	.997	.997	.064
Reliability (alpha)	.999	.999	-47.318

are almost equal to one (theoretically they should be one) (Table 9.9).

Table 9.10 *Squared Item-Total Correlation*

Comp.	Item 1	Item 2	Item 3	Item 4	Item 5	Item 6
1	.442	.336	1.000	.559	.121	.065
2	.353	.072	1.000	.037	.198	.062
3	.656	.602	.000	.140	.507	.588

The criterion item has a perfect squared correlation with Components 1 and 2, and nil correlation with the third component, which is a conditional component (Table 9.10). The eigenvalues for the proper components (adjusted eigenvalues) can be calculated by the formula mentioned earlier as

$$\rho*_1{}^2 = \frac{0.442 + 0.336 + 1.000 + 0.559 + 0.121 + 0.065 - 1.000}{5}$$
$$= 0.305$$

$$\rho*_2{}^2 = \frac{0.353 + 0.072 + 1.000 + 0.037 + 0.198 + 0.062 - 1.000}{5}$$
$$= 0.144$$

Since the variable 'age' contributes to the total proper space by two and the sum of the two proper eigenvalues is 0.449 (=0.305+0.144), we can state that the *commonality* of the criterion item 'age' is given by the percentage of 0.449 to 2.000, which is 22.5 percent. In other words, 22.5 percent of 'age' can be explained by the other five items and the remaining 77.5 percent is its unique contribution to the total space.

The commonality of 22.5 percent does not sound very high. Our question then is how well we can predict which option of the criterion variable each subject has chosen from his or her responses to the five non-criterion items. Let us look at scores of subjects on the two proper components (Table 9.11).

Let us plot subjects, using these component scores as coordinates and label them in terms of the options of the criterion variable (Age) they have chosen (Figure 9.1). In spite of the impression one gets from 22.5 percent, the plot shows that we can predict the age categories 100 percent correct! It is easy to introduce boundaries between age categories in the graph to attain this 100 percent prediction.

Table 9.11 *Chosen Options of Age and Scores*

Subj	Option	Comp.1	Comp.2
1	3	-.48	-.06
2	1	-.32	.73
3	3	-.45	-.21
4	3	-.57	-.32
5	2	.98	-.07
6	2	.61	-.15
7	2	.94	-.19
8	1	-.14	.58
9	2	.92	-.28
10	2	.46	.24
11	1	.16	.55
12	3	-.02	-.10
13	3	-.44	-.40
14	1	.01	.53
15	3	-.47	-.31

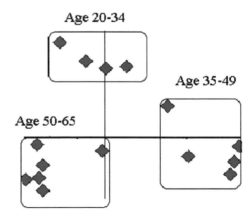

Figure 9.1 *Forced Classification on Age*

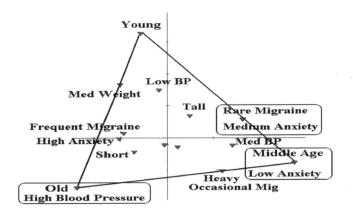

Figure 9.2 *Forced Classification on Age*

We can also look at an interesting configuration of options of the six items in this forced classification analysis. Figure 9.2 shows the plot of the options using their weights of the proper components as coordinates. Those in boxes in the graph are the options which occupy the same or almost the same coordinates. From this graph, we can see that options of the non-criterion variables all fall within the triangle created by the options of the criterion variable. This is so because in forced classification non-criterion variables are projected onto the space of the criterion variables, that is, three options of the age variable. Two unlabeled points near the center of the graph are ones which do not contribute to the interpretation of the results. The age group 'old' coincides exactly with 'high blood pressure.' The same can be said about middle age and low anxiety, and rare migraine and medium anxiety. The cluster of frequent migraines, short height and high anxiety are close to the line connecting the old age group and the young group, but tends to be closer to the old age group. Heavy weight is close to low anxiety and middle age group.

In forced classification, the criterion variable is perfectly correlated with the total score, and the item-total correlation is identical to the item-criterion correlation. For instance, let us look at the inter-item correlation matrix for Component 1 (Table 9.12). Look at the correlation coefficients

Table 9.12 *Inter-Item Correlation for Component 1*

Item 1	1.00					
2	0.86	1.00				
3	0.66	0.58	1.00			
4	0.45	0.32	0.75	1.00		
5	-0.10	0.09	0.35	0.15	1.00	
6	0.22	0.15	0.26	0.23	-0.44	1.00
Item	1	2	3	4	5	6

of five items with Item 3, that is, 0.66, 0.58, 0.75, 0.35 and 0.26, respectively. These are equal to the square roots of the squared item-total correlation coefficients in the first row of Table 9.10.earlier table.

Component 3 of the current forced classification is the first component after partialing out the effect of the variable 'age.' Its eigenvalue is the average of the remaining five values in the table, which is

$$\rho_1{}^2(5|age) = \frac{0.656 + 0.602 + 0.000 + 0.140 + 0.507 + 0.588}{5}$$

$$= 0.499$$

Note that the criterion variable no longer has any more contributions to Component 3 and the remaining components, that is, conditional components. Therefore, forced classification offers a procedure to eliminate the effects of the criterion variable such as age from analysis, and allows us to see results free from the effect of the criterion variable.

It may be possible that a survey sample of chosen cosmetics, for example, may be dominated by young subjects, but that the cosmetics company may be seeking results applicable to all ages. In this type of situation, one may use Age as the criterion for forced classification, and look at only conditional components.

As we have seen, forced classification is a procedure to use the options of the criterion variable to determine principal axes. As such, it is the procedure to derive maximal correlation between the criterion variable and each of the remaining variables. Thus no other scoring

Table 9.13 *Sorting of Animals by the Ideal Sorter*

Pile 1	Pile 2	Pile 3	Pile 4
frog	crow	dog	alligator
lizard	pigeon	chimpanzee	cheetah
tortoise	chicken	cow	bear
snake	duck	cat	tiger
	sparrow	rabbit	leopard
	turkey	monkey	lion
	crane	goat	rhinoceros
	hawk	pig	giraffe
	ostrich	horse	hippopotamus
		racoon	elephant
		fox	camel

scheme would provide a higher average correlation coefficient between the criterion item and the rest of the items than forced classification.

9.4 Ideal Sorter of Animals

Let us look at an application of forced classification to sorting data. Recall that sorting data can be regarded as transposed multiple-choice data. Thus, instead of choosing a criterion item, we will choose a criterion subject, and see if we can predict from responses of non-criterion subjects the piles that the criterion subject has used, or see which of the non-criterion subjects show sorting patterns similar to those of the criterion subject. In forced classification of animals, Mayenga (1997) chose an ideal sorter as the criterion subject. In his case, the ideal sorter was one who classified animals following the well-known Linnaean classification system.

In our example, let us use the sorting data as discussed earlier, that is, Tables 8.2 and 8.3. We will augment the data in Table 8.3 with one more subject's responses, say the ideal subject's classification of animals, and use this subject as the criterion. Suppose that this criterion subject classified animals in Table 8.2 as in Table 9.13, that is, into four groups.

Let us augment the data in Table 8.3 by adding this sorter as Subject 16 in the last column of the table (Table 9.14). Let us carry out forced classification with Subject 16 as the criterion. Since this subject used four

Table 9.14 *Classification of Animals Augmented by Ideal Sorter 16*

Ss	1	2	3	4	5	6	7	8	9	10	11	12	13	14	15	16
A0	1	1	1	1	1	1	1	1	1	1	1	1	1	1	1	3
A1	2	2	2	2	2	2	3	3	2	2	2	2	3	2	8	4
A2	3	1	3	3	3	1	4	3	3	3	3	2	3	2	3	3
A3	4	1	4	4	2	3	4	1	2	1	1	1	1	3	4	2
A4	5	3	5	5	5	4	2	5	4	4	4	4	4	6	5	2
A5	5	3	5	5	5	4	2	5	4	4	4	1	4	6	5	2
A6	6	1	1	6	1	2	3	4	5	1	3	1	2	4	2	4
A7	5	3	5	10	5	4	6	5	4	4	1	4	4	6	5	2
A8	1	1	6	4	2	2	3	3	5	5	5	1	1	4	2	4
A9	1	1	1	6	1	2	1	1	1	1	1	1	2	1	1	3
B0	1	1	7	4	1	5	7	1	1	1	1	1	5	1	1	3
B1	2	2	8	9	6	3	5	2	2	2	6	3	6	7	6	1
B2	6	1	4	4	4	5	7	4	1	1	1	1	5	1	1	3
B3	3	1	1	6	1	2	3	4	5	1	2	1	2	4	2	4
B4	7	1	3	4	2	5	3	4	5	2	5	1	7	5	2	4
B5	4	1	9	7	4	5	3	4	5	3	5	1	7	5	4	4
B6	5	3	5	10	5	4	2	5	4	4	4	4	4	6	5	2
B7	5	3	5	5	5	4	2	5	4	3	4	4	4	6	5	2
B8	7	1	2	4	2	5	3	2	5	2	5	1	7	5	2	4
B9	1	1	3	3	3	1	4	3	3	3	3	2	3	2	3	3
C0	5	3	5	10	5	4	6	5	4	4	1	4	4	6	5	2
C1	4	1	4	4	4	2	7	4	1	1	1	1	5	3	1	3
C2	5	3	5	5	5	4	2	5	4	4	4	4	4	6	5	2
C3	4	1	1	6	1	2	3	4	5	1	2	1	2	4	2	4
C4	6	3	5	10	5	4	3	5	4	4	5	4	4	5	4	2
C5	8	2	8	9	6	3	5	2	2	2	6	3	8	7	6	1
C6	1	1	4	4	4	1	3	4	1	1	1	1	5	3	4	3
C7	1	1	1	1	0	2	7	1	1	1	2	1	1	1	1	3
C8	2	2	8	2	6	3	5	2	2	5	6	3	6	7	6	1
C9	2	2	8	2	2	3	5	2	2	2	6	3	8	7	6	1
D0	3	1	1	6	2	2	3	4	5	1	2	1	7	4	2	4
D1	7	1	2	4	4	5	3	4	5	5	2	1	7	5	2	4
D2	7	1	4	7	4	5	3	4	5	3	5	1	7	5	4	4
D3	5	3	5	5	5	4	2	5	4	4	2	4	4	6	5	2
D4	1	1	1	6	1	2	7	1	1	1	2	1	1	1	1	3

Table 9.15 *Adjusted Information of Proper Components*

	Component 1	Component 2	Component 3
Eigenvalue	.88	.79	.45
Singular Value	.94	.89	.67
Adjusted Delta	29.3	26.3	15.0
Cum.delta	29.3	55.6	70.6

piles, there are three proper components. The total information of the proper components is as in Table 9.15. These eigenvalues and singular values are adjusted ones, and adjusted delta values are calculated by

$$\text{Adjusted Delta} = \frac{\text{Adjusted Eigenvalue} \times 100}{m_c - 1} \qquad (9.10)$$

where m_c is the number of piles that the criterion subject used. The large

Table 9.16 *Correlation between Criterion Subject and Other Subjects*

Subject	1	2	3	4	5	6	7	8
Component 1	.91	.90	.99	.92	.90	.99	.99	.90
Component 2	.87	.90	.94	.91	.90	.94	.88	.91
Component 3	.66	.01	.59	.61	.54	.53	.88	.58

Subject	9	10	11	12	13	14	15	
Component 1	.91	.66	.98	.98	.99	.99	.99	
Component 2	.90	.84	.69	.87	.94	.89	.92	
Component 3	.95	.38	.68	.20	.76	.96	.93	

value of the cumulative delta, 70.6, indicates that the criterion subject's sorting data share a great deal of responses with the rest of the group. Let us see this aspect of the results through the correlation between the criterion subject and the non-criterion subjects over the proper components (Table 9.16).

It is amazing to see high correlation coefficients of those non-criterion subjects with Subject 16, in particular in Components 1 and 2. Thus we can anticipate a plot of these two components to be very similar to the one obtained earlier in the standard MUNDA run (Figure 8.3). Let us look at the plot of animals using Component 1 and Component 2 (Figure 9.3). We can see that two of the piles that the criterion subject (Subject 16) used are distinctly clustered in Figure 9.3, suggesting that we need another dimension to see the other two piles. Thus, we must look at Component 3. Figure 9.4 is the plot of Component 3 against Component 1.

Component 3 indeed identifies the remaining two piles created by the criterion subject. Thus, the sorting responses of the criterion subject (Subject 16) can be fully described in three-dimensional space. Although a three-dimensional graph is not presented here, one can visualize it, and it is clear that the four piles of the criterion subject can be correctly clustered in three-dimensional space with 100 percent accuracy. In other words, using those weights of 15 non-criterion subjects, we can identify the memberships of the animals in the four groups employed by the criterion subject, that is, we can correctly predict the classification by the criterion subject. For instance:

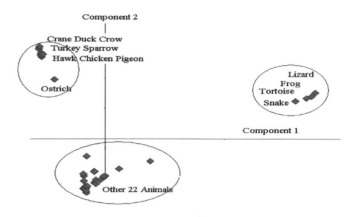

Figure 9.3 *Forced Classification on Subject 16, Components 1 and 2*

1. If the score of Component 1 is greater than 2.0, the animal belongs to Pile 1.
2. If the score of Component 1 is negative and that of Component 2 is greater than 0.9, the animal belongs to Pile 2.
3. If the score of Component 1 is negative and that of Component 3 is positive, the animal belongs to Pile 3.
4. If the score of Component 1 is less than 0.6 and that of Component 3 is negative, the animal belongs to Pile 4.

With this decision tree, the classification of animals is 100 per cent correct.

We have looked at two applications of forced classification, one to multiple-choice data and the other to sorting data. Unlike the standard MUNDA, forced classification allows us to focus our attention to a single variable of interest. As such, it has a great potential for applications to practical problems such as market segmentation problems (e.g., Can we predict first choices of cars by consumers' ages?) and investigation of rare phenomena that may not surface under the

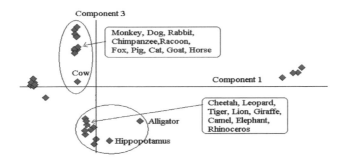

Figure 9.4 *Forced Classification on Subject 16, Components 1 and 3*

standard analysis (e.g., Is the behavior of being startled at telephone ringing related to any neurological disorders?).

9.5 Generalized Forced Classification

The procedure described so far was generalized to a number of situations (Nishisato, 1986a). It was with respect to multiple-choice data in the response-pattern format $\mathbf{F}$ with N rows and m columns, partitioned into n submatrices $\mathbf{F}_j$ of $N \times m_j$. Let $\mathbf{P}$ and $\mathbf{Q}_j$ be respectively $N \times N$ and $m_j \times m_j$ diagonal matrices of weights, that is, $\mathbf{P}=\mathrm{diag}(p_i)$, $i=1,2,...,N$ and $\mathbf{Q}_j=\mathrm{diag}(q_{js})$, $s=1,2,...,\ m_j$ and $j=1,2,\ ...,n$. Consider the following general expression for forced classification,

$$\mathbf{F}(p_i, q_{js}) = (\mathbf{PF}_1\mathbf{Q}_1, \mathbf{PF}_2\mathbf{Q}_2, \cdots, \mathbf{PF}_j\mathbf{Q}_j, \cdots, \mathbf{PF}_n\ \mathbf{Q}_n) \quad (9.11)$$

The generalization can take many different forms, and Nishisato (1986a) provided only the following examples:

1. The standard forced classification is carried out by setting that $\mathbf{P}=\mathbf{I}$, $\mathbf{Q}_j=k\mathbf{I}_j$, and all other matrices $\mathbf{Q}_p$ are identity matrices of respective orders.

2. When we suspect 'outliers' as was the case with Peter Tung's data, use the weight of 0.5 for the elements of $\mathbf{Q}_j$, that is, $\mathbf{Q}_j=0.5\mathbf{I}$

corresponding to exceptionally small frequencies and for the elements of **P** of subjects with exceptionally rare responses, that is, **P** is the identity matrix, except for a few diagonal elements which are set equal to 0.5. Gibson (1993) explored the applicability of generalized forced classification for successfully dealing with outlier responses.

3. In personality testing, we sometimes want to identify which subjects belong to a particular personality type. Then, augment the data matrix by including a subject (ideal subject) who responds to the questionnaire with a typical response pattern (e.g., Type A personality), and carry out forced classification with a large constant k for the corresponding element of **P**. Day (1989) successfully used this version of generalized forced classification to sort over 2300 English speaking subjects and over 2300 French speaking subjects into sixteen personality types, using the Myers-Briggs Type Indicator (Myers, 1962).

4. We can use more than one item as the criterion item, for example, by specifying $\mathbf{Q}_j = k\mathbf{I}_j$ and $\mathbf{Q}_{j+1} = k\mathbf{I}_{j+1}$ with an extremely large value of k.

5. We can specify more than one subject as the criterion, for example, by setting $p_i = k$ and $p_{i+1} = k$, with a large enough a value of k.

6. What will happen if we substitute optimal option weights for all the elements of $\mathbf{Q}_j$, $j = 1, 2, \cdots, n$? In this question, one might ask what the meaning of using negative weights means.

7. What will happen if we substitute optimal scores of subjects for the elements of **P**? The same question as above about the use of negative scores as weights.

8. What will happen if we substitute optimal option weights and optimal scores for the elements of $\mathbf{Q}_j$ and **P**? Nishisato (1988c) considered dual scaling of this general expression, and discussed the reversed procedure of transforming the input data matrix, calling it 'inverse dual scaling.' The final results, however, have never been published. The inverse quantification has also been investigated by Groenen and van de Velden (2004), and interested readers are referred to their paper.

When we specify more than one item or more than one subject as the criteria for forced classification, we are focusing our attention to a larger subset of the data matrix than in the standard forced classification. Thus, for a subset of the data to become dominant, the forcing agents (values of k) must be larger than the case of a one criterion variable. In addition, the number of proper forced classification components increases accordingly.

One should note, too, that the nature of quantification changes quite drastically by this generalization. When the forcing agents are taken to infinity, the case can be interpreted in terms of projecting the data set onto the subspace jointly occupied by the set of criterion variables, be it a subgroup of subjects or a subset of items. Otherwise, the notion of projection becomes unclear. For instance, when we use the weight of 0.5 for two options, how can we interpret the quantification process as an analytical method of the original data? As the value of k gets smaller, the results of analysis approaches to that of the data matrix projected by the operator $(\mathbf{I}-\mathbf{P})$. If we use optimal weights for generalized forced classification, it looks as though we are looking at the data in principal space, defined by the original data. Then, what will be the results of generalized forced classification? In all these forms of generalization , is it still appropriate to seek such a statistic as *delta* and use it to interpret the analysis? Generalized forced classification is a very important and interesting topic since it opens a new area of research. At the current moment, however, applications have been limited to the cases where more than one variable are used as criteria and large enough values of k are used for the analysis. Forced classification as a suppressor of outlier responses needs further investigation.

PART III

Analysis of Dominance Data

Dominance data are characterized by such responses as 'greater than,' 'equal to,' or 'less than.' Since the information is typically inequality, without any specific amount of the discrepancy between the two attributes or objects, it is not possible to approximate the ordinal information directly as is done with the incidence data. To handle ordinal measurement in a "purely nonmetric" way, that is, as ordinal numbers, however, creates almost insoluble problems for multidimensional quantification. Therefore, we treat ordinal numbers as cardinal and adopt the least-squares approach to the quantification problem of dominance data. With this treatment of dominance data as cardinal numbers, we will see that we no longer have to deal with the chi-square metric, so long as missing responses are not involved, but we can use the Euclidean metric, which simplifies a number of otherwise tricky problems. In evaluating the outcome of quantification, however, we must use a special objective such that the ranking of the derived measurements best approximates the corresponding ranking of the original dominance data. Since we treat ordinal measurement as cardinal, we are essentially predicting the data elements as was the case of incidence data, but the evaluation of the scaling outcome must be done in a way suitable for ordinal measurement.

In Part 2, we discussed (1) contingency tables, (2) multiple-choice data and (3) sorting data. Included in Part 3 are dominance data of the following types:

(4) Paired comparison data
(5) Rank order data
(6) Successive categories data

We will discuss quantification of each of these data types. Recall the characteristics of dominance data as summarized in 4.2.2.

CHAPTER 10

Paired Comparison Data

10.1 Example

Paired comparison data consist of an $N \times [n(n-1)/2]$ (subjects-by-pairs of stimuli, $X(j,k)$) table, of which the element in the i-th row and the (j,k)-th column is 1 if Subject i prefers Stimulus j to Stimulus k, 0 if Subject i makes an equality judgment, and 2 if Subject i prefers Stimulus k to Stimulus j. The following example shows 3 subjects who made paired comparison judgments of four fruits, apples (A), pears (P), grapes (G) and mangos (M). The columns correspond to the pairs (A, P), (A, G), (A, M), (P, G), (P, M), and (G, M).

We should note that in paired comparisons it is possible for a

Table 10.1 *Paired Comparison Data*

Subject	AP	AG	AM	PG	PM	GM
1	1	2	2	2	2	2
2	2	1	1	2	1	2
3	1	2	2	1	2	2

subject to produce intransitive relations such as the subject prefers A to B, B to C, and C to A, rather than A to C. Such intransitive judgments must arise, for example, from subject's adopting different judgmental criteria for different pairs, such as sweetness, spiciness, flavor and nutritional value. In the unidimensional scaling (Bock and Jones, 1968), intransitive judgments create some problems since they are considered to be associated with errors. In multidimensional quantification, however, intransitive judgments present interesting and perhaps important pieces of information.

10.2 Early Work

Guttman's monumental work on paired comparison and rank-order data was published in 1946. Some of the later work in the 1960s and 1970s by other researchers were essentially the same as Guttman's methods, although superficially they look different. Thus, let us trace some historical developments.

10.2.1 Guttman's Formulation

Unlike quantification of incidence data, scaling of paired comparison data requires some coding. Guttman proposed the following coding. For a pair of two objects (X_j, X_k) judged by Subject i, define

$$_ie_{jk} = \begin{cases} 1 & \text{if Subject i judges} \quad X_j > X_k \\ 0 & \text{if Subject i judges} \quad X_j < X_k \end{cases} \tag{10.1}$$

$i = 1, 2, ..., N; j, k = 1, 2, ..., n, (j \neq k)$. Define two variables f_{ij} and g_{ij} as follows,

$$f_{ij} = \sum_{k=1}^{n} {}_ie_{jk} \qquad g_{ij} = \sum_{k=1}^{n} {}_ie_{kj} \tag{10.2}$$

f_{ij} is the total number of judgments in which Subject i preferred X_j to the remaining objects and g_{ij} is that in which Subject i preferred other objects to X_j. Note the sum of these variables is always equal to n-1 for every i and every j. Historically, it is interesting, as well as important, to notice that Guttman's use of these quantities is essentially identical to the concept of "doubling" in correspondence analysis, discussed by Greenacre (1984).

To determine the weight vector for n objects in such a way that the objects are maximally discriminated, Guttman (1946) solved the following eigenequation,

$$(\mathbf{H}_g - \lambda \mathbf{I})\mathbf{x} = \mathbf{0}, \tag{10.3}$$

where

$$\mathbf{H}_g = \frac{2}{Nn(n-1)^2}(\mathbf{F'F} + \mathbf{G'G}) - \frac{1}{n}\mathbf{1}_n\mathbf{1}_n' \tag{10.4}$$

and $\mathbf{F} = (f_{ij})$ and $\mathbf{G} = (g_{ij})$. Guttman's procedure does not deal with negative numbers, and in this regard it complies with the condition of correspondence analysis that any data can be subjected to correspondence analysis so long as data elements are non-negative (Greenacre, 1984).

10.2.2 Nishisato's Formulation

Nishisato (1978, see also 1994) presented an alternative formulation to Guttman's. He introduced the following response variable, including one for a tied rank:

$$
{}_i f_{jk} = \begin{cases} 1 & \text{if Subject } i \text{ judges} \quad X_j > X_k \\ 0 & \text{if Subject } i \text{ judges} \quad X_j = X_k \\ -1 & \text{if Subject } i \text{ judges} \quad X_j < X_k \end{cases} \tag{10.5}
$$

$i=1,2,...,N;$ $j,k=1,2,...,n$ $(j \neq k)$. In terms of his coding, the example in Table 10.1 can be expressed as in Table 10.2. He defines the *dominance*

Table 10.2 *Coded Paired Comparison Data*

Subject	AP	AG	AM	PG	PM	GM
1	1	-1	-1	-1	-1	-1
2	-1	1	1	-1	1	-1
3	1	-1	-1	1	-1	-1

number for Subject i and Object j by

$$
e_{ij} = \sum_{k=1}^{n} {}_i f_{jk} \tag{10.6}
$$

In applying this formula, we must define that

$$
{}_i f_{jk} = -{}_i f_{kj} \tag{10.7}
$$

For instance,

$$
\begin{aligned}
e_{1A} &= {}_1 f_{AP} + {}_1 f_{AG} + {}_1 f_{AM} \\
&= 1 - 1 - 1 = -1 \\
e_{1P} &= {}_1 f_{AP} + {}_1 f_{PG} + {}_1 f_{PM} = -{}_1 f_{PA} + {}_1 f_{PG} + {}_1 f_{PM} \\
&= -1 - 1 - 1 = -3 \\
e_{1G} &= {}_1 f_{AG} + {}_1 f_{PG} + {}_1 f_{GM} = -{}_1 f_{GA} - {}_1 f_{GP} + {}_1 f_{GM} \\
&= 1 + 1 - 1 = 1 \\
e_{1M} &= {}_1 f_{AM} + {}_1 f_{PM} + {}_1 f_{GM} = -{}_1 f_{MA} - {}_1 f_{MP} - {}_1 f_{MG} \\
&= 1 + 1 + 1 = 3
\end{aligned}
$$

The judge-by-object table of e_{ij} is called "dominance matrix" and is indicated by $\mathbf{E}$. The task of quantification is to determine weights for subjects so as to maximize the variance of scale values of objects, where scale values are defined as the differentially weighted averages of subjects' dominance numbers. In Nishisato's formulation, the eigen-equation is given by

$$(\mathbf{H}_n - \lambda\mathbf{I})\mathbf{x} = \mathbf{0}, \tag{10.8}$$

where

$$\mathbf{H}_n = \frac{1}{Nn(n-1)^2}\mathbf{E}'\mathbf{E} \tag{10.9}$$

Recall Guttman's definition of two matrices $\mathbf{F}$ and $\mathbf{G}$, and note that Nishisato's matrix $\mathbf{E} = \mathbf{F} - \mathbf{G}$. Then, after a few simple algebraic manipulations, Nishisato reached the conclusion that $\mathbf{H}_n = \mathbf{H}_g$. In other words, his alternative formulation is mathematically equivalent to Guttman's.

Furthermore, Nishisato (1994, 1996) showed that we can map the subjects and objects in multidimensional space in such a way that each subject will choose the object closer to him or her than the other object by plotting normed weights of subjects and projected weights of objects. This was the main object of quantifying rank order data (i.e., Coombs' unfolding method, to be discussed in Chapter 11), and we now know that the same results apply to paired comparison data as well.

10.3 Some Basics

As for the formulation of the quantification procedure, Nishisato's 1978 method deals with the dominance matrix $\mathbf{E}$. The element (i,j) of the dominance matrix, e_{ij} can be calculated from Nishisato's coding variable $_if_{jk}$ as discussed earlier with a numerical example. If that formula is difficult to use, one can obtain the dominance matrix from the coded variables and the design matrix of paired comparisons (Bock and Jones, 1968). Let the $N \times n(n-1)/2$ coded matrix be indicated by $\mathbf{E}^*$ where the columns are arranged in the specific order, for example, (A,B), (A,C), (A,D), (B,C),(B,D), (C,D) for a set of four objects (A,B,C,D). Let $\mathbf{A}$ be the design matrix $\mathbf{A}$. Then, the dominance matrix $\mathbf{E}$ is given by

$$\mathbf{E} = \mathbf{E}^*\mathbf{A} \tag{10.10}$$

Table 10.3 *Dominance Table*

Fruit	Apple	Pear	Grapes	Mango	Sum
1	-1	-3	1	3	0
2	1	1	-1	-1	0
3	-1	-1	-1	3	0

If we apply this method to our numerical example,

$$
\mathbf{E} = \begin{bmatrix} 1 & -1 & -1 & -1 & -1 & -1 \\ -1 & 1 & 1 & -1 & 1 & -1 \\ 1 & -1 & -1 & 1 & -1 & -1 \end{bmatrix} \begin{bmatrix} 1 & -1 & 0 & 0 \\ 1 & 0 & -1 & 0 \\ 1 & 0 & 0 & -1 \\ 0 & 1 & -1 & 0 \\ 0 & 1 & 0 & -1 \\ 0 & 0 & 1 & -1 \end{bmatrix}
$$

$$
= \begin{bmatrix} -1 & -3 & 1 & 3 \\ 1 & 1 & -1 & -1 \\ -1 & -1 & -1 & 3 \end{bmatrix}
$$

Thus, the dominance table is as given in Table 10.3. Each dominance number indicates the number of times the fruit is preferred to the others minus the number of times the other fruits were preferred to the fruit.

If we look at each fruit and its dominance numbers from three subjects, we can see the extent of individual differences. If we sum them over subjects, we can roughly tell which fruit was most preferred by the group. In our quantification, however, we would like to quantify individual differences, rather than averaging them out. Thus, we assign weights to the subjects in such a way that the weighted averages of fruits attain the maximal variance.

Let us note a special aspect of the dominance table. The sum of each row is zero. Thus, when we analyze $\mathbf{E}$, we cannot use the row marginals or column marginals of $\mathbf{E}$ as indicators of the numbers of responses. Nishisato (1978) rationalized the counting of the numbers in the following way. When a subject judges, for example, apple, he or she compares apple with the remaining three fruits. Thus, the element in the dominance table associated with apple and this subject is the result of three comparisons. In this way, Nishisato (1978) proposed that each element of the dominance matrix is based on n-1 comparisons. Therefore each row of the dominance table (i.e., the row marginal in the incidence data) is based on $n(n-1)$ responses, and each column on $N(n-1)$ responses, hence

$f_t = Nn(n-1)$. In matrix notation,

$$\mathbf{D}_r = (n-1)\mathbf{I}, \quad \mathbf{D}_c = N(n-1)\mathbf{I} \qquad (10.11)$$

With these new definitions of the marginals, the rest is the eigenvalue decomposition of the dominance matrix $\mathbf{E}$, or more precisely that of matrix $\mathbf{H}_n$ of (10.9).

When the paired comparison data are converted to the dominance table $\mathbf{E}$, the table size is reduced from $N \times n(n-1)/2$ to $N \times n$. Noting that $\mathbf{E1=0}$, the rank of the dominance table is $n-1$, provided that N is larger than $n-1$. We have also stated that dominance data do not have a trivial solution. Thus, the total number of components, $T(comp)$, is given by

$$T(comp) = n - 1, \quad \text{proivided that } N \geq n - 1$$
$$T(comp) = N, \quad \text{otherwise} \qquad (10.12)$$

As for the sum of eigenvalues, Nishisato (1993) derived it for both paired comparison data and rank-order data. Since both have the same expression of $T(inf)$ and its derivation for rank-order data is simpler than for paired comparison, see Chapter 11. $T(inf)$ for both data types is

$$T(inf) = \sum_{k=1}^{n-1} \rho_k{}^2 = \frac{n+1}{3(n-1)} \qquad (10.13)$$

Notice that the sum of eigenvalues attains its maximum of 1 when $n=2$ and decreases to $\frac{1}{3}$ when n goes to infinity. Thus, $T(inf)$ is bounded between these two values. This has two implications. The first one is a welcome situation related to the statistic δ_k. This statistic is defined as before, that is, the percentage of the ratio of the eigenvalue of the component to the sum of all the eigenvalues. We now have the situation in which the first eigenvalue can be as large as the sum of the eigenvalues, meaning the possibility that the first component can account for 100 percent of the information in the data. Recall the case of the contingency table, in which we noted that when the table is greater than 2×2, the first component cannot account for 100 percent of information in the data. The second implication of the $T(inf)$ is that even when one component accounts for 100 percent of the information in data, ρ^2 can not be 1, unless the number of objects is 2; for a larger number of objects, ρ^2 can never reach 1.

10.4 Travel Destinations

Data were collected from students when the author was teaching at a university in Nishinomiya, Japan. Twenty students were asked with respect to each pair of travel destinations which one in the pair they would prefer visiting. The eight destinations were used: Athens (A), Honolulu (H), Istanbul (I), London (L), Madrid (M), Paris (P), Rome (R) and Sydney (S), which created 28 pairs: (A,H), (A,I), (A,L), (A,M), (A,P), (A,R), (A,S), (H,I), (H,L), (H,M), (H,P), (H,R), (H,S), (I,L), (I,M), (I,P), (I,R), (I,S), (L,M), (L,P), (L,R), (L,S), (M,P), (M,R), (M,S), (P,R), (P,S), (R,S). These pairs were randomized and presented to the students for pair comparisons. The responses are coded as 1 if the first destination in a pair was preferred to the second one and 2 if the second one was preferred to the first one (Table 10.4).

Let us look at the dominance table to see how those eight cities are viewed by the subjects (Table 10.5). As we recall, the dominance number of a subject indicates the number of times the person has chosen a particular city over other cities minus the number of times the person chose other cities over the city. So, let us look at Athens (A). 7 means that A was preferred to other seven cities; 5 means A was preferred to other six cities and one city was preferred to A (6-1=5); 3 means A was preferred to five other cities and two cities were preferred to A (5-2=3); 1 means A was preferred to four other cities and three cities were preferred to A (4-3=1); -1 means A was preferred to three other cities and four cities were preferred to A (3-4=-1); -3 means A was preferred to two other cities and five cities were preferred to A (2-5=-3); -5 means that A was preferred to one city and the remaining six cities were preferred to A (1-6=-5), and ; -7 means the other seven cities were preferred to A (0-7=-7).

It is amazing to notice that there are no cities which show consistent dominance numbers over subjects, that is, all positive numbers or negative numbers. This suggests that there are substantial individual differences in preference for travel destinations, reflecting perhaps such factors that they may have visited some of these cities before, or some familiarity to particular cities, or that they may have adopted different criteria such as exotic cities, ancient cities, glamorous cities or modern cities.

The information is distributed over seven components as in Table 10.6, which shows two dominant components. Notice that because the eigenvalues are bounded between 0 and $\frac{n+1}{3(n-1)}$, they are much smaller than those we observed for incidence data. From the third component, the

Table 10.4 *Paired Comparisons of Travel Destinations*

| X_j | A | A | A | A | A | A | A | H | H | H | H | H | H | I |
X_k	H	I	L	M	P	R	S	I	L	M	P	R	S	L
1	1	2	2	2	1	2	1	2	2	2	2	2	2	2
2	1	1	1	1	2	1	1	2	2	2	2	2	2	2
3	1	1	2	1	2	2	2	1	1	1	2	2	2	2
4	1	1	2	1	2	2	1	1	2	1	1	2	2	2
5	2	1	2	2	2	2	2	1	2	2	2	1	2	2
6	1	1	1	1	1	1	1	2	2	2	2	2	2	1
7	2	1	1	1	1	2	1	1	2	1	1	1	1	1
8	1	1	2	1	2	2	1	2	2	2	2	2	1	2
9	1	2	1	2	1	1	1	2	2	2	2	2	1	1
10	1	1	1	2	1	1	1	2	2	2	2	2	1	1
11	1	1	1	2	1	1	1	2	2	2	2	2	1	1
12	1	1	1	1	1	2	1	2	2	2	2	2	2	1
13	1	1	1	1	1	1	1	2	2	2	2	2	2	1
14	2	2	2	2	2	2	2	1	1	1	2	1	1	2
15	1	2	1	1	1	1	2	2	1	1	1	2	1	1
16	1	2	1	1	1	1	1	2	1	1	1	1	1	1
17	2	2	2	2	2	2	2	1	1	1	1	1	1	1
18	2	2	2	2	2	2	2	1	1	1	1	1	2	2
19	2	2	2	2	2	2	2	2	2	2	2	2	2	2
20	1	1	1	1	1	1	1	2	2	2	2	2	2	1

| X_j | I | I | I | I | L | L | L | L | M | M | M | P | P | R |
X_k	M	P	R	S	M	P	R	S	P	R	S	R	S	S
1	2	1	1	1	1	1	1	1	1	1	1	2	1	1
2	2	2	2	1	1	2	2	1	2	2	1	1	1	1
3	1	2	2	2	1	2	1	2	2	2	1	1	1	2
4	2	2	2	2	1	1	1	1	2	2	2	2	1	1
5	2	2	2	2	1	2	1	1	2	1	1	1	1	2
6	1	1	1	1	1	2	1	1	1	1	1	1	1	1
7	1	2	2	2	1	2	2	2	2	1	1	1	2	2
8	2	2	2	1	1	2	1	1	2	1	1	1	1	1
9	2	2	1	1	2	1	2	1	1	1	1	2	1	1
10	1	1	2	1	1	1	2	2	2	1	1	2	1	1
11	1	1	2	1	1	1	2	2	2	1	1	2	1	1
12	2	2	2	1	1	1	2	1	1	2	1	2	1	1
13	2	1	2	1	2	1	2	2	1	2	1	2	2	1
14	1	2	2	1	1	1	2	1	2	2	1	2	1	1
15	1	1	1	2	1	2	2	2	2	2	2	1	2	2
16	1	1	1	1	2	2	2	1	2	2	1	2	1	1
17	1	1	1	1	1	2	1	2	2	2	2	2	2	2
18	2	2	2	2	1	1	1	2	2	2	2	2	2	2
19	2	2	2	2	1	1	1	1	2	2	2	1	1	2
20	1	1	1	1	2	2	2	2	2	2	2	1	2	2

Table 10.5 *Dominance Numbers of Eight Cities by Subjects*

	Athens	Honolulu	Istanbul	London	Madrid	Paris	Rome	Sydney
1	-1	-7	3	7	5	-3	1	-5
2	5	-7	-3	1	-1	5	5	-5
3	-1	-1	-5	1	-5	7	1	3
4	1	-1	-7	7	-5	1	5	-1
5	-5	-1	-7	5	3	7	-3	1
6	7	-7	5	1	1	1	-3	-5
7	3	5	-3	-3	-3	1	-1	1
8	1	-5	-3	5	1	7	1	-7
9	3	-5	3	-1	7	-1	1	-7
10	5	-5	3	-1	1	-1	3	-5
11	5	-5	3	-1	1	-1	3	-5
12	5	-7	-1	1	1	-1	7	-5
13	7	-7	1	-3	3	-5	5	-1
14	-7	5	-1	3	-3	3	5	-5
15	3	1	5	-5	-7	-1	-1	5
16	5	3	7	-5	-3	-1	1	-7
17	-7	7	5	-1	-5	-1	-1	3
18	-7	5	-5	3	-3	-1	1	7
19	-7	-5	-3	7	-1	5	1	3
20	7	-7	5	-5	-3	1	-1	3

Table 10.6 *Distribution of Information over Seven Components*

Component	1	2	3	4	5	6	7
Eigenvalue	.15	.10	.04	.03	.02	.01	.01
Singular Value	.39	.31	.20	.18	.15	.14	.08
Delta	41.2	26.2	11.0	8.7	6.2	5.0	1.8
Cum.delta	41.3	67.4	78.4	87.1	93.3	98.2	100.

information reduction is rather gradual, although there is a sharp drop from component 6 to component 7. This may be a typical case in which researchers decide to adopt only two components because of the two dominant components and the ease of interpretation. However, it is often tricky to decide the number of components only by looking at the distribution of information, for even a minor component may be of some crucial importance to researchers (e.g., when one is looking for a rare symptom of illness).

In interpreting the quantification results, we should remember that for dominance data we plot the normed weights of subjects and the projected weights of cities. When this choice is made, the plot has the follow-

Table 10.7 *Normed Weights for Subjects in Three Components*

S	1	2	3	4	5	6	7	8	9	10
C.1	-.91	-1.04	.58	.20	.74	-1.42	.37	-.61	-1.33	-1.23
C.2	1.02	1.17	.79	1.35	1.62	.02	-.52	1.68	.26	.02
C.3	1.73	-1.44	=1.96	-1.31	.06	-.08	-1.03	-.27	1.32	-.06

S	11	12	13	14	15	16	17	18	19	20
C.1	1.23	-1.29	-1.40	.69	.09	-.83	1.15	1.55	.62	-.97
C.2	.02	.77	-.22	.76	-1.49	-1.23	-.98	.28	1.62	-.87
C.3	-.06	-.52	-.37	.58	-1.50	.09	.54	-.23	-.29	-1.73

S=Subject; C=Component

Table 10.8 *Projected Weights for Cities in Three Components*

City	A	H	I	L	M	P	R	S
C.1	-.64	.61	-.30	.13	-.24	.11	-.17	.49
C.2	-.20	-.32	-.43	.51	.14	.32	.17	-.20
C.3	-.24	.11	.17	.11	.35	-.21	-.09	-.20

C=Component

ing meaning: provided that it is a good representation of the data each subject is located closer to the city she or he prefers than to the other city not preferred to in the pair. The coordinates of subjects, that is, their normed weights, and the coordinates of cities, that is, their projected weights for the first three components are given in Table 10.7 and Table 10.8, respectively.

What will happen if we look at only the first component, ignoring

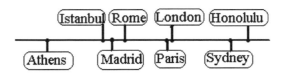

Figure 10.1 *Travel Destinations in Space of Component 1*

others? See Figure 10.1. This graph reflects 41 per cent of the total information. Thus, this unidimensional scale does not represent everyone's preference values of those travel destinations. We should therefore look at the configuration of the first two components (Figure 10.2). It is interesting to note that the configuration shows four clusters of two cities each which are located close to each other. Thus, those who choose, for example, London are likely to choose Paris as well. Similar arguments seem to hold for other clusters (Rome,Madrid), (Honolulu,Sydney) and (Athens,Istanbul). For interpretation of data, it is important for us to look at both cities and subjects, for the city configuration is based on the subject configuration, and vice versa.

Figure 10.3 and Figure 10.4 are joint plots of subjects and cities, based

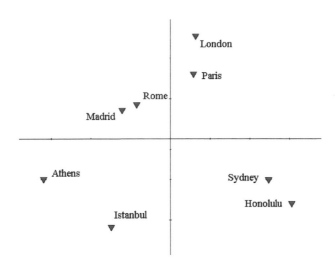

Figure 10.2 *Travel Destinations in Space of Components 1 and 2*

on Component 1 against 2 and Component 1 against 3, respectively. Cities are indicated by triangles and subjects by squares. Since cities are projected onto the subject space, the cities are located closer to the origin than the subjects.

In Figure 10.3, the closest cities for some subjects or clusters of subjects

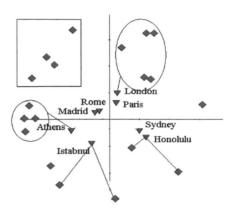

Figure 10.3 *Joint Display of Components 1 and 2*

are indicated. If those first choices are also the same as in the original
data, then those points are satisfactory from the recovery point of view.
We can see that several subjects are closest to London and then to Paris
(their first and second choices), a few are closest to Athens and then to
Madrid and Rome. Figure 10.3 also shows subjects who are closest to
Honolulu and those closest to Istanbul. What is the most important point
here is the fact that subjects are distributed all over the space. If a unidimensional model was correct, all the subjects would be located on the
horizontal axis at the distance of 1 from the origin. This is definitely
not the case here. For those subjects who are in a rectangular area, they
are distributed almost in parallel to the line along which London, Paris,
Rome, Madrid and Athens are located.

Figure 10.4 is presented here to show that the joint configuration of
dominance data is quite different from that of incidence data, for as
in Figure 10.3 the subjects' configuration is normed or standardized,
irrespective of dimension. In Figure 10.3 and Figure 10.4, look at the

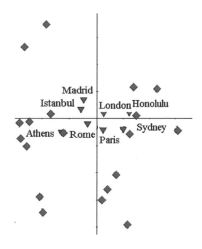

Figure 10.4 *Joint Display of Components 1 and 3*

vertical dispersion of subjects, in contrast to the small dispersion of cities.

10.5 Criminal Acts

Data were collected in Nishisato's scaling class in 1979. Nineteen students participated in paired comparison judgment of six criminal acts:
  (A) Arson, (B) Burglary, (F) Forgery,
  (H) Homicide, (K) Kidnaping, (M) Mugging.
There are in total 15 pairs, that is, 6(6-1)/2. In this data set, we see a number of tied responses, coded 0. For example, Subject 13 could not distinguish in seriousness between arson and homicide, arson and kidnaping, arson and mugging, burglary and forgery, burglary and mugging and kidnaping and mugging.

Although there may be exceptional cases, we can safely expect some consistency over subjects that homicide, arson and kidnaping are more serious than burglary and forgery. Let us look at dominance numbers

Table 10.9 *Paired Comparisons of Six Criminal Acts*

X_j	A	A	A	A	A	B	B	B	B	F	F	F	H	H	K
X_k	B	F	H	K	M	F	H	K	M	H	K	M	K	M	M
1	1	1	2	2	2	1	2	2	2	2	2	2	1	1	1
2	1	1	2	0	1	1	2	2	0	2	2	2	1	1	1
3	1	1	2	2	1	1	2	2	1	2	2	1	1	1	1
4	1	1	2	2	1	1	2	2	2	2	2	2	1	1	2
5	1	1	2	2	1	1	2	2	2	2	2	2	1	1	0
6	1	1	2	1	1	1	2	2	2	2	2	2	1	1	1
7	1	1	2	2	1	1	2	2	1	2	2	2	1	1	1
8	0	1	2	2	2	1	2	2	1	2	2	2	1	1	2
9	1	1	2	2	2	1	2	2	2	2	2	2	1	2	1
10	1	1	0	1	1	1	2	2	0	2	2	2	1	1	0
11	1	1	0	1	1	1	2	2	2	2	2	2	1	1	1
12	1	1	2	0	1	1	2	2	2	2	2	2	1	1	1
13	1	1	0	0	0	0	2	2	0	2	2	2	1	1	0
14	1	1	2	0	1	1	2	2	0	2	2	2	1	1	1
15	1	1	2	1	2	1	2	2	2	2	2	2	1	1	2
16	1	1	1	1	0	0	2	2	0	2	2	0	1	1	1
17	0	1	2	2	2	0	2	2	2	2	2	2	1	1	1
18	1	1	1	1	1	2	2	2	2	2	2	1	1	1	1
19	1	1	2	1	1	1	2	2	2	2	2	2	1	1	1

Note on crimes: 1=A;2=B;3=C;4=F;5=H;6=K;7=M;8=R

(Table 10.10) As expected, dominance numbers of homicide are consistently positive and those of forgery consistently negative. This is a good indication that the data must contain a dominant first component.

The judgment of seriousness of criminal acts is very much in line with what we expect, which is reflected in the distribution of δ values over five components (Table 10.11).

As is clear, the major portion of information is explained by the first component. Therefore, let us first look at the first component (Figure 10.5). This scale corresponds to what most people look at the seriousness of those criminal acts because the first component explains 84 percent of total information. Figure 10.6 shows the joint plot of Component 1 against Component 2. Notice that the second component's contribution to the locations of the criminal acts is relatively small, while the locations of subjects on the second component is large, due to the fact that we plotted the normed weights of the subjects. As expected, homicide is the most serious criminal act, followed by kidnaping, arson, mugging, burglary and forgery in this order. The fact that Component 1 explains 84 percent of information indicates this order of seriousness is mostly accepted by this group of subjects.

There are two strange aspects of these two plots. The first aspect is

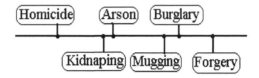

Figure 10.5 *Criminal Acts on Component 1*

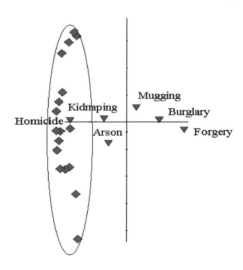

Figure 10.6 *Component 1 against Component 2 of Criminal Acts*

Table 10.10 *Dominance Table of Six Criminal Acts*

Crime	A	B	F	H	K	M
Subject 1	-1	-3	-5	5	3	1
2	2	-2	-5	5	2	-2
3	1	-1	-3	5	3	-5
4	1	-3	-5	5	1	1
5	1	-3	-5	5	2	0
6	1	-3	-5	5	3	-1
7	1	-1	-5	5	3	-3
8	-2	0	-5	5	1	1
9	-1	-3	-5	3	3	3
10	4	-2	-5	4	0	-1
11	4	-3	-5	4	1	-1
12	2	-3	-5	5	2	-1
13	2	-3	-4	4	1	0
14	2	-2	-5	5	2	-2
15	1	-3	-5	5	1	1
16	4	-3	-3	3	1	-2
17	-2	-3	-4	5	3	1
18	5	-5	-1	3	1	-3
19	3	-3	-5	5	1	-1

Table 10.11 *Distribution of Information over Components*

	C1	C2	C3	C4	C5
Eigenvalue	.35	.04	.02	.01	.00
Singular Value	.60	.20	.13	.09	.05
Delta	84.1	9.7	4.0	1.8	p.5
Cum. delta	84.1	93.8	97.7	99.5	100.

the fact that Component 2 also shows a wide spread of subjects. This is so because we use normed weights of subjects for all dimensions. But, since most subjects are close to the location -1 from the origin on the first component, this is a typical two dimensional graph where the first component is dominant. Most researchers would be satisfied by just looking at Component 1 to interpret the data.

The second strange aspect is an important one for MUNDA, namely

Table 10.12 *Normed Weights of Subjects in Two Components*

Subject	1	2	3	4	5	6	7
Component 1	-1.05	-1.09	-.92	-1.03	-1.08	-1.13	-1.06
Component 2	1.26	-.35	-.84	.55	.37	.20	-.18

Subject	8	9	10	11	12	13	14
Component 1	-.81	-.85	-.98	-1.06	-1.12	-.91	-1.09
Component 2	1.58	1.66	-.88	-.86	-.16	-.12	-.35

Subject	15	16	17	18	19
Component 1	-1.03	-.82	-.94	-.79	-1.12
Component 2	.55	-1.33	1.47	-2.15	-.52

Table 10.13 *Projected Weights of Criminal Acts in Two Components*

Crime	A	B	F	H	K	M
Component 1	-.30	.51	.90	-.91	-.36	.15
Component 2	-.37	.06	-.12	.05	.08	.29

the direction of an axis is arbitrary. In the second example, one would normally expect that "Homicide" should receive the largest value of seriousness. Our example showed -0.91 for "Homicide," that is, the smallest value. Thus, if this happens and if one does not like the direction of the values, we may simply change the sign of all the weights, which would still be optimal quantities, that is, weights that maximize, for example, the correlation ratio. This is very much like looking at Niagara Falls from the American side and then from the Canadian side: the location of the falls change the directions, for instance, the one on the left-hand side changes to the right-hand side.

In this chapter, we observed two distinct examples, one in which many individual differences are involved and the other with very little individual differences. The joint plots are quite different in these two extreme cases. When we see in practice a graph which is similar to the second example, we should know that the first component is highly dominant and perhaps sufficient to explain the nature of data.

Since we use normed weights for subjects and projected weights for objects (e.g., travel destinations, criminal acts), both quantities occupy the same space. In our case, we project objects onto the space of subjects. Thus, we can calculate the distance between a subject and an object. As we discussed earlier, we should calculate the distance between Subject i and Object j, d_{ij}, the distance between Subject i and Object k, d_{ik}. If Subject i prefers Object j to Object k in the given data and d_{ij} is greater than d_{ik} for all i, j and k, then we can say that it is a perfect solution.

This rank order comparison of distances in reduced space (e.g., in two-dimensional space) with that of the original data will be discussed in Chapter 11, to evaluate how good an approximation to the input data we have in terms of a few components.

Rank-Order Data

In Chapter 10, we noted that Guttman (1946) proposed a procedure for quantifying both paired comparison data and rank-order data. We also saw that Nishisato's 1978 alternative formulation was identical to Guttman's and can handle rank-order data as well.

Thus, we do not have to change the paired comparison formulation to handle rank-order data, except that as we will see later it can be simplified because of the nature of rank-order data. Thus, most of the previous discussion applies to rank-order data, including $T(comp)$, $T(inf)$, dominance numbers, the specification of the number of responses in each cell of the dominance table and its theoretical marginals.

11.1 Example

Rank-order data consist of a table of ranks of n objects by N subjects, and its i-th row and the j-th column contains the rank number given to Stimulus j by Subject i. The following is an example in which 3 subjects ranked four movies, with 1 indicating the most preferred and 4 the least.

In the social sciences, we often talk about *ipsative data*, characterized

Table 11.1 *Rank-Order Data*

Movie	Pepe	Gigi	Tom	Anna
1	2	3	4	1
2	2	1	4	3
3	1	3	2	4

by the fact that elements in each row (or column) sum up to a constant for all rows (columns). It is well known that when the data are row-ipsative, the between-column correlation is meaningless. For example, if you consider the ranking of two cars, A and B, by 100 consumers and if we decide to treat rank 1 and rank 2 as cardinal numbers, then the product-moment correlation between A and B is always -1. Similarly, if we ask the same people to rank five cars and treat rank numbers as cardinal, then the correlation between any two cars is likely to be negative, because the ipsative nature of the data is inherited in the tendency of one variable getting a higher rank or a lower rank at the expense of the other variables. The rank-order data arranged in the above example are also referred to as *row-conditional data*, which implies that the comparisons of rank numbers are meaningful only within each row. Weingarden and Nishisato (1986) demonstrated that paired comparison data capture more individual differences than rank order data when the same stimuli were judged by the two methods, paired comparisons and ranking. This is understandable because ranking is always transitive while paired comparison can be intransitive (i.e., A beats B, B beats C and C beats A).

11.2 Early Work

When we talk about rank order data, what comes to our mind first is Coombs' multidimensional unfolding method, which began with his paper on a unidimensional unfolding model in 1950. It was a new type of scaling technique for at least two reasons: that it takes into consideration individual differences, and that it was based on purely nonmetric procedure. Coombs (1950) postulated a quantitative scale, called J scale, along which both objects (to be ranked) and subjects are located. His model is simple: each subject ranks first the object (e.g., a political party) closest to him or her, ranks second the object next closest to the person, and so on. In Coombs' model, each subject has his or her "ideal point" (or position, scale value) on the continuum, and when the continuum is folded at the person's ideal point so that objects now lie on the folded continuum, we can produce the person's ranking of the objects according to the order in which the objects appear on the folded continuum. Coombs considered that the rank-order data are rankings of objects along this folded continuum. Thus, his task was to unfold a set of rank-orders from subjects to recover the original J continuum, that is, the locations of subjects and objects. Since different subjects are likely to have different ideal points (locations on the J scale), folding the J scale at different points will generate many rank-orders of the stimuli. We should also note that the folded J scale can provide only a limited number of rankings of stimuli.

It is interesting to note, however, that Guttman (1967) complained about Coombs' lack of attention to Guttman's earlier work (Guttman,

1941), and in particular the work on scaling of rank order data and paired comparison data (Guttman, 1946). Why did Coombs (1950) ignore Guttman's contributions? It was probably because Guttman's work was based on treating ordinal measurement as cardinal numbers while Coombs' approach was strictly nonmetric and Coombs must have believed in the strict procedure of handling ordinal measurement as is, that is, only a monotone transformation is permissible for ordinal measurement.

When Coombs' unidimensional unfolding model was presented in 1950, his scaling problem was solved by linear programming. An easier method than linear programming was also proposed by Goode (1957). Phillips (1971) further elaborated on Goode's method as a method for solving a set of linear inequality equations.

Coombs' original unidimensional unfolding model soon encountered a problem: real data provided more rank-orders than his unidimensional model could generate. If we draw a line and place objects and subjects, we can then fold the line at each subject's location, each time producing one rank-order. It is easy to figure out that a unidimensional model can produce only a limited number of rankings out of $n!$ possible combinations of n objects. Thus, the model was extended to a multi-dimensional model in which objects and subjects are distributed in multidimensional space, with the same idea for the generation of rank order: each subject ranks first the closest object, second the second closest object and so on, the difference now being that both subjects and objects lie in multidimensional space (see, for example, Coombs, 1964; Bennett, 1956; Bennett and Hays, 1960; Hays and Bennett, 1961).

In this multidimensional extension, however, there was no explanation of what "folded scale" would look like. Nevertheless, the idea of "unfolding" was applied to the multidimensional extention. The extension made it possible to deal with real data. However, it encountered a tricky problem of solving his multidimensional unfolding problem in an ordinal way. In as early as in 1960, Coombs and Kao (1960) appeared to have abandoned a nonmetric approach, moving towards treating ordinal measurement as cardinal numbers. The same was the case with Ross and Cliff (1960).

When Schönemann (1970) published his famous paper on metric multidimensional unfolding, it was clear that the original intention of solving an ordinal problem with ordinal measurement was completely abandoned. By treating ordinal measurement (rank order 1, 2, 3, and so on) as cardinal numbers (integer 1, 2, 3, and so on), it became possible to adopt the least-squares approach to solving the scaling problem.

After Schönemann's monumental work, a large number of papers on multidimensional unfolding were published, to name a few, Schönemann and Wang (1972), Davidson (1973), Gold (1973), Sixtl (1973), Heiser (1981), Greenacre and Browne (1986) and Adachi (2000).

Thus, in retrospect, Guttman's complaint (1967) about Coombs' ignoring Guttman's papers is well justified. Guttman (1946), rather than Coombs (1950), was the person who laid the foundation for multi-dimensional quantification of rank-order data. Guttman's formulation handles both rank-order data and paired comparison data, while Coombs' unfolding method deals only with rank-order data. Apart from Guttman's complaint, we should perhaps add another complaint: Researchers in the area of multidimensional unfolding often ignore contributions by Guttman and others in the field of quantification approach, considering in particular that Guttman's approach provides a solution to Coombs' problem of multidimensional unfolding of rank orders.

Along Guttman's line of approach, a number of studies were published, to name a few, Tucker (1960), Slater (1960), Carroll (1972), de Leeuw (1973), Nishisato (1978), Skinner and Sheu (1982), Hojo (1994), Han and Huh (1995) and Okamoto (1995). Out of these, Nishisato (1978) demonstrated mathematical equivalence of methods by Slater (1960), Tucker (1960) and Carroll (1972) to Guttman's, and presented an alternative formulation to Guttman's, extending it to handle tied judgments in rank order and paired comparisons. In addition to those mentioned so far, Baba (1986) presented a very different but interesting approach to the scaling of rank orders, and it is worth reading.

In 1994, Nishisato discovered that the Coombs' multidimensional un-folding problem could be solved by dual scaling if we plot subjects' normed weights and objects' projected weights, that is, if we project objects onto subjects' space. His treatment of ordinal measurement as cardinal makes it very practical in handling a large data set. In contrast, a nonmetric approach would encounter difficulty when a data set becomes as large as most rank-order data we see in practice. The researchers, therefore, have moved away, at least for the time being, from Coombs' original idea of "pure" nonmetric approach to Guttman's cardinal approach to analysis of ordinal measurement.

11.3 Some Basics

In Chapter 10, we saw that Guttman's approach yields the following eigenequation,

$$(\mathbf{H}_g - \lambda \mathbf{I})\mathbf{x} = \mathbf{0}, \tag{11.1}$$

where

$$\mathbf{H}_g = \frac{2}{Nn(n-1)^2}(\mathbf{F'F} + \mathbf{G'G}) - \frac{1}{n}\mathbf{1}_n\mathbf{1}_n' \tag{11.2}$$

$\mathbf{F} = (f_{ij})$ and $\mathbf{G} = (g_{ij})$, where

$$f_{ij} = \sum_{k=1}^{n} {}_ie_{jk} \qquad g_{ij} = \sum_{k=1}^{n} {}_ie_{kj} \tag{11.3}$$

and

$$_ie_{jk} = \begin{cases} 1 & \text{if Subject i judges } X_j > X_k \\ 0 & \text{if Subject i judges } X_j < X_k \end{cases} \tag{11.4}$$

$i = 1, 2, ..., N; j, k = 1, 2, ..., n, (j \neq k)$.

In Chapter 10, we saw Nishisato's approach, yielding the following eigenequation,

$$(\mathbf{H}_n - \lambda\mathbf{I})\mathbf{x} = \mathbf{0}, \tag{11.5}$$

where

$$\mathbf{H}_n = \frac{1}{Nn(n-1)^2}\mathbf{E'E} \tag{11.6}$$

and elements of $\mathbf{E}$ is obtained as follows: Define

$$_if_{jk} = \begin{cases} 1 & \text{if Subject i judges } X_j > X_k \\ 0 & \text{if Subject i judges } X_j = X_k \\ -1 & \text{if Subject i judges } X_j < X_k \end{cases} \tag{11.7}$$

$i=1,2,...,N; j,k=1,2,...,n$ $(j \neq k)$, and the dominance number for Subject i and Object j by

$$e_{ij} = \sum_{k=1}^{n} {}_if_{jk} \tag{11.8}$$

or, using the familiar design matrix for paired comparisons, $\mathbf{A}$ (see Bock and Jones, 1968), the matrix of dominance numbers is obtained by $\mathbf{E}=\mathbf{FA}$, where $\mathbf{F}$ is an $N \times n(n-1)/2$ matrix of $_if_{jk}$.

In 1973, de Leeuw proposed the following formula for rank order data,

$$e_{ij} = n + 1 - 2R_{ij} \tag{11.9}$$

where R_{ij} is the rank that Subject i gave to Object j out of n objects. Nishisato (1978) showed that his formula for dominance numbers for rank-order data can also be expressed simply by de Leeuw's formula.

In Chapter 10, we noted that Guttman's two matrices $\mathbf{F}$ and $\mathbf{G}$ are related to the dominance matrix: $\mathbf{E} = \mathbf{F} - \mathbf{G}$, hence that $\mathbf{H}_n = \mathbf{H}_g$.

Nishisato (1978) also established the equivalence of the Guttman-Nishisato approach to other studies, to be presented next.

11.3.1 Slater's Formulation

Slater's formulation is based on the matrix $\mathbf{S}$, which is the row-centered matrix of Guttman's $\mathbf{F}$, that is,

$$\mathbf{S} = \mathbf{F} - \frac{(n-1)}{2} \mathbf{1_N 1_n'} \qquad (11.10)$$

and the eigenequation to be solved is given by

$$(\mathbf{H}_s - \lambda \mathbf{I})\mathbf{x} = \mathbf{0} \qquad (11.11)$$

where $\mathbf{H}_s = \mathbf{SS'}$. Thus, this is principal component analysis of the row-centered matrix of $\mathbf{F}$, that is, $\mathbf{S}$. In terms of Guttman's $\mathbf{F}$ and $\mathbf{G}$ and Nishisato's $\mathbf{E}$, we can express Slater's matrix $\mathbf{S}$ as follows,

$$\mathbf{S} = \mathbf{F} - \frac{1}{2}(\mathbf{F} + \mathbf{G}) = \frac{1}{2}(\mathbf{F} - \mathbf{G}) = \frac{1}{2}\mathbf{E} \qquad (11.12)$$

Thus, we can see that Slater's formulation is also equivalent to Guttman's and Nishisato's formulations (Nishisato, 1978).

Slater (1960) talks about principal component analysis (PCA) of rank-order data. Noting the fact that each element of the dominance matrix is based on a constant number of comparisons, that is, n-1, we can use Euclidean metric, not chi-square metric. This underlies the equivalence between PCA and the quantification approach. If missing responses are involved, the above equivalence disappears. Slater's approach reminds us of Torgerson's nomenclature for the quantification method as "principal component analysis of categorical data."

11.3.2 Tucker-Carroll's Formulation

The formulations by Tucker (1960) and Carroll (1972) can be described, using Guttman's notation as follows. Define

$$\mathbf{E_{tc}} = \sqrt{w_i} \sum_{j \neq k}^{n} (_i e_{jk} - _i e_{kj}) \qquad (11.13)$$

where w was introduced to differentially weight subjects. When $w = 1$, $\mathbf{E_{tc}}$ is the same as Nishisato's $\mathbf{E}$, and the eigenequation is formulated for the product of this matrix, thus leading to the same conclusion that their formulations, too, are equivalent to the others discussed above (Nishisato, 1978).

11.4 Total Information and Number of Components

Rank-order data are transformed to the $N \times n$ dominance table $\mathbf{E}$, which is row-conditional. Thus, the total number of components, $T(comp)$, is

$$T(comp) = n - 1, \quad \text{if } N \geq n - 1$$
$$T(comp) = N, \quad \text{otherwise} \tag{11.14}$$

As for the information contained in the $N \times n$ dominance table, it is given by the trace of matrix $\mathbf{H}_n$,

$$T(inf) = \sum \rho_k{}^2 = trace(\mathbf{H_n}) = \frac{1}{Nn(n-1)^2} trace(\mathbf{E}'\mathbf{E}) \tag{11.15}$$

Noting that the elements of $\mathbf{E}$ can be generated by the de Leeuw formula, we can obtain the trace of $\mathbf{E}'\mathbf{E}$ as

$$
\begin{aligned}
tr(\mathbf{E}'\mathbf{E}) &= tr(\mathbf{EE}') = N \sum_{j=1}^{n}(n+1-2R_j)^2 \\
&= N \sum ((n+1)^2 - 4(n+1)R_j + 4R_j{}^2) \\
&= Nn(n+1)^2 - 4N(n+1)\sum R_j + 4N \sum R_j{}^2 \tag{11.16} \\
&= Nn(n+1)^2 - 4N(n+1)\frac{n(n+1)}{2} + \frac{4Nn(n+1)(2n+1)}{6} \\
&= \frac{Nn}{3}(n-1)(n+1)
\end{aligned}
$$

Therefore, the trace of $\mathbf{H_n}$ is given by

$$T(inf) = tr(\mathbf{H_n}) = \frac{1}{Nn(n-1)^2} tr(\mathbf{E}'\mathbf{E}) = \frac{n+1}{3(n-1)} \tag{11.17}$$

Thus, the total information is bounded by

$$\frac{1}{3} \leq T(inf) \leq 1 \tag{11.18}$$

$T(inf)$ becomes a minimum when n goes to infinity and the maximum of 1 when $n=2$.

11.5 Distribution of Information

The last formula for the total information tells us an interesting fact that the total information is solely a function of the number of objects and is independent of the number of judges, a characteristic stemming from the row-conditionality of the data matrix. This has a number of implications for the distribution of information in rank-order data. Let us start with special cases.

11.5.1 The Case of One Judge

The independence of the total information of the number of judges seems to imply that our quantification method has no numerical difficulty in solving the eigenequation when there is only one judge! The data set then consists of a string of numbers, and a single component exhaustively explains the entire data, irrespective of the number of objects. Although one solution exhaustively explains the total information, the eigenvalue varies between one third and one, depending on the number of objects.

Thus, quantification of data from many judges then is only a matter of calculating differentially weighted configurations of objects coming from individual judges.

11.5.2 One-Dimensional Rank Order Data

There are three obvious cases when data can be explained by one component, first when there is only one judge as we saw in the previous section, second when there are only two objects, irrespective of the number of judges, and third when all the judges provide the same ranking of objects, irrespective of the number of objects and the number of judges. These are special cases, and serve to show some important differences between Coombs' approach and the quantification approach.

11.5.3 Coomb's Unfolding and MUNDA

It may be obvious by now that Coombs' approach and our approach MUNDA are different. First of all, when data are collected from a single judge, Coombs' method cannot provide a quantitative scale as a single set of ranks does not generally provide enough quantitative information. Similarly, when our quantification method provides a single component that exhaustively explains the entire information in the data, Coombs' method fails to provide any quantitative scale.

The concept of dimensionality is a point of departure between them. In our approach, the rank numbers are treated as if they were real numbers of equal units, and as such the dimensionality is defined by the rank of the dominance matrix. In contrast, Coombs' approach is an ordinal approach in which one can distort a multidimensional configuration in any way one likes so long as ordinal information is retained. In this regard, Coombs' approach is logically more sound than our approach as a technique to handle ordinal data.

Let us look at a numerical example where Coombs' approach provides a better result than MUNDA. If one is familiar with Coombs' unidimen-

Table 11.2 *Coombs' Unidimensional Data*

Object						
Judge	A	B	C	D	E	F
1	1	2	3	4	5	6
2	2	1	3	4	5	6
3	2	3	1	4	5	6
4	2	3	4	1	5	6
5	3	2	4	1	5	6
6	3	4	2	1	5	6
7	3	4	2	5	1	6
8	3	4	5	2	1	6
9	3	4	5	2	6	1
10	4	3	5	2	6	1
11	4	3	5	6	2	1
12	4	5	3	6	2	1
13	5	4	3	6	2	1
14	5	4	6	3	2	1
15	5	6	4	3	2	1
16	6	5	4	3	2	1

sional unfolding model, it is easy to tell that the above data set in Table 11.2 will yield a unidimensional J scale. Notice that Coombs' condition of a unidimensional scale is met: only one pair of adjacent ranks changes as we go down the table from Judge 1 to Judge 16 when rank orders are appropriately arranged.

In contrast, we need more than one dimension if we analyze the same data with our approach, more specifically it requires five dimensions for a complete description of the data. The five delta values are as in Table 11.3. This example clearly shows the difference between Coombs'

Table 11.3 *Five Delta Values by Our Approach*

Component	1	2	3	4	5
Delta	55.45	22.86	12.66	6.74	2.26

approach and MUNDA, for the latter handles rank orders as cardinal numbers and the rank of our sample data is five.

Thus, a disadvantage of MUNDA over Coombs' unfolding analysis lies in the extra dimensions needed to accommodate all rank orders in the data set. In addition, we have the problem of discrepant spaces, that is, the row space and the column space, which does not arise in Coombs' approach.

Most importantly, however, MUNDA can handle data collected from any number of judges, even as few as one judge or as many as ten thousand judges, while Coombs' approach often encounters the problem of insufficient information in the data or too much information to deal with. Thus, from the practical point of view, any of the Guttman-type methods has a great advantage over Coombs' approach. It should be noted, too, that Coombs' method often faces "ill-conditioned data" where a closed solution is not available. This absence of a solution is not a problem with MUNDA.

Finally, MUNDA offers a solution to Coombs' multidimensional unfolding problem (Nishisato, 1994, 1996) when normed weights for subjects and projected weights for objects are jointly plotted in the same space, that is, when objects are projected onto the subject space. The joint configuration contains the information for each subject as to which object is to be chosen first, which one for second, and so on for all objects and for all subjects. Should some one solve Coombs' problem in a purely nonmetric way for a very large data set, it is possible that Coombs' joint configuration may be accommodated in the space of fewer dimensions than the configuration by MUNDA. But, how much fewer? The discrepancy in the number of dimensions between the two approaches may be trivial, however, since we hardly look at more than several dimensions.

11.5.4 Goodness of Fit

The traditional statistic of δ, "the percentage of the total information explained," is useful in many cases.

$$\delta_j = 100 \frac{\lambda_j}{T(inf)} \tag{11.19}$$

where λ_j is the j-th eigenvalue of $\mathbf{H_n}$. However, since we are dealing with rank orders and our objective is to reproduce input rank orders in the space of the smaller dimensions than the data set requires, a better statistic than the above is desirable.

Following Nishisato's paper (1994), we first plot normed weights of N subjects and n projected weights of objects in k-dimensional space ($k=1,2,..., K$), compute the Euclidean distance between Subject i ($i=1,2,..., N$) and Object j ($j= 1,2,...,n$),

$$d_{ij(k)} = \sqrt{\sum_{p=1}^{k}(y_{i(p)} - \rho_p x_{j(p)})^2} \qquad (11.20)$$

rank these distances from the smallest (to be designated as rank 1) to the largest (rank n), and call these ranks of the distances as the order-k approximations of Subject i's input (original) ranks. Let us indicate by R^*_{ij}
Subject i's recovered rank of Object j.

Nishisato (1996) proposed two statistics: (1) the sum of squares of rank discrepancies between observed ranks and recovered ranks for each component, or multiple components, for each judge or all the judges, and; (2) the percentage of the above statistic as compared to the worst ranking, that is, the reversed ranking of the observed for each judge ($\delta_{ij}(rank)$) or all the judges ($\delta_j(rank)$). From the practical point of view, the first one is more useful than the second one, and we will use the first one, say $D^2(i : k)$, and its average over n objects, to eliminate the effect of the number of objects. These discrepancy measures of Subject i for Dimension k are

$$D^2(i : k) = \sum_{j=1}^{n}(R_{ij} - R*_{ij})^2 \qquad (11.21)$$

$$\overline{D^2}(i : k) = \frac{D^2(i : k)}{n} \qquad (11.22)$$

The use of these statistics will be illustrated in the next numerical example.

11.6 Sales Points of Hot Springs

The data were collected when the author was teaching at a university in Nishinomiya, Japan. One of the favorite pastimes of Japanese is to relax at hot springs, which are scattered over most parts of this volcanic country. Twenty-four adults were asked to rank the following items according to the preference order of why they choose their destination hot spring:

(A) A day trip possible (Day)
(B) Less than 10,000 yen (roughly 100 dollars) a night (10,000yen)
(C) Outdoor hot spring bath available (Outdoor)
(D) Family bath, rather than public, available (Family)
(E) Beautiful scenery surrounding it (Scenery)
(F) Known for excellent food (Food)
(G) Sports/exercise facilities available (Sports)
(H) Located in a resort hotel (Resort)
(I) Recreational facilities available (Recreation)
(J) Pets allowed to stay (Pet)

Rank 1 is the most preferred and rank n the least preferred out of n objects, hot spring attractions in the above example. Table 11.4 shows the ranking of 10 sales points of hot springs by 24 subjects. In the case

Table 11.4 *Ranking of Ten Hot Springs' Sales Points*

	A	B	C	D	E	F	G	H	I	J
1	9	7	6	8	1	5	3	2	4	10
2	9	10	7	8	3	6	2	4	5	1
3	8	10	4	9	2	6	3	5	7	1
4	9	4	2	1	5	3	7	6	8	10
5	6	1	3	7	5	2	4	6	8	10
6	10	9	1	5	2	3	6	4	7	8
7	9	2	1	5	4	3	8	6	7	10
8	7	5	3	9	4	1	8	6	2	10
9	5	1	6	7	3	2	9	8	4	10
10	8	3	1	9	4	2	5	7	6	10
11	5	3	6	9	2	1	8	7	4	10
12	9	8	6	5	4	1	2	3	7	10
13	8	9	4	7	2	1	6	3	5	10
14	6	4	9	5	3	1	2	10	7	8
15	9	10	6	2	4	5	8	3	7	1
16	8	7	1	10	2	3	6	4	5	9
17	9	8	5	4	3	2	10	6	7	1
18	5	1	6	7	3	2	9	8	4	10
19	7	6	8	3	4	5	1	9	10	2
20	1	10	6	9	4	5	8	3	7	2
21	10	5	1	8	2	3	6	4	7	9
22	10	9	1	5	2	3	4	6	7	8
23	9	1	2	8	3	4	7	6	5	10
24	10	7	6	2	4	5	3	8	9	1

A=Day; B=10,000yen; C=Outdoor; D=Family; E=Scenery;
F=Food; G=Sports; H=Resort; I=Recreation; J=Pet

of rank-order data, the dominance table can easily be obtained by the de Leeuw formula as in Table 11.5. One of the differences between

Table 11.5 *Dominance Table*

	A	B	C	D	E	F	G	H	I	J
1	-7	-3	-1	-5	9	1	5	7	3	-9
2	-7	-9	-3	-5	5	-1	7	3	1	9
3	-5	-9	3	-7	7	-1	5	1	-3	9
4	-7	3	7	9	1	5	-3	-1	-5	-9
5	-1	9	5	-3	1	7	3	-1	-5	-9
6	-9	-7	9	1	7	5	-1	3	-3	-5
7	-7	7	9	1	3	5	-5	-1	-3	-9
8	-3	1	5	-7	3	9	-5	-1	7	-9
9	1	9	-1	-3	5	7	-7	-5	3	-9
10	-5	5	9	-7	3	7	1	-3	-1	-9
11	1	5	-1	-7	7	9	-5	-3	3	-9
12	-7	-5	-1	1	3	9	7	5	-3	-9
13	-5	-7	3	-3	7	9	-1	5	1	-9
14	-1	3	-7	1	5	9	7	-9	-3	-5
15	-7	-9	-1	7	3	1	-5	5	-3	9
16	-5	-3	9	-9	7	5	-1	3	1	-7
17	-7	-5	1	3	5	7	-9	-1	-3	9
18	1	9	-1	-3	5	7	-7	-5	3	-9
19	-3	-1	-5	5	3	1	9	-7	-9	7
20	9	-9	-1	-7	3	1	-5	5	-3	7
21	-9	1	9	-5	7	5	-1	3	-3	-7
22	-9	-7	9	1	7	5	3	-1	-3	-5
23	-7	9	7	-5	5	3	-3	-1	1	-9
24	-9	-3	-1	7	3	1	5	-5	-7	9

A=Day; B=10,000yen; C=Outdoor; D=Family; E=Scenery;
F=Food; G=Sports; H=Resort; I=Recreation; J=Pet

incidence data and dominance data is that dominance data must first be converted to the table of dominance numbers. If there were no individual differences, the reasonable scale values or appealing values of the ten hot spring attributes would be given by the average dominance numbers over subjects. However, we assume that individual differences are worthwhile variables, rather than random fluctuations from the mean. The scale values of the ten sales points are calculated as differentially weighted averages. Our main task is to determine weights for subjects such that the variance of the weighted means of hot spring attractions be a maximum. We should remember that individual differences create multidimensional data structure.

T(comp) is 9 and all nine values of δ are shown in Table 11.6. Considering a relatively sharp drop from Component 2 to 3, let us examine the first two components. As mentioned earlier, there is a

Table 11.6 *Distribution of Information*

Component	1	2	3	4	5
δ	41.10	23.05	11.86	8.13	6.72
Cum. δ	41.10	64.15	76.01	84.13	90.85

Component	6	7	8	9
delta	4.46	2.56	1.28	0.84
Cum.δ	95.32	97.88	99.16	100.0

special plotting rule for dominance data: Plot normed weights of subjects and projected weights of objects. This joint plot of the first two components is as shown in Figure 11.1. To see how good this joint configuration is, we first calculate the distance between each sales point and each subject from this two-dimensional configuration. Table 11.7 shows the subjects-by-sales points table of squared distances. We now convert these squared distances to ranks, the smallest as 1, the second smallest as 2 and so on, within each subject. The resultant rank-order table is called the rank-2 approximation to the original data. The rank-2 approximations are shown in Table 11.8. The last column of Table 11.8 lists the sum of squared discrepancies between the original ranks and rank-2 approximations over ten sales points. This statistic tells us how good an approximation by Figure 11.1 is to the original data. The smallest discrepancy is indicated by Subject 6 with the sum of squared discrepancies of 4. Table 11.9 shows this subject's original ranking and rank-2 approximation to the ten ranks. Look at Figure 11.1, based on the first two components. Inverted triangles indicate ten sales points, and squares show subjects. Notice also that since we are projecting ten sales points onto the space of subjects, the former ten points tend to be closer to the origin.

Since the rank-2 approximation to Subject 6 is excellent, we indicated in Figure 11.1 the position of Subject 6 and also the ranks of Scenery (1), Resort (4) and Pet (8) with arrows to show their distances from Subject 6. In the data of this subject, rank 1 was Outdoor, but in this approximation Outdoor is ranked 2, meaning some additional components are needed to rectify this discrepancy. Resort and Pet are correctly approximated. It is important to note that if we look at the configuration of sales points, it may be difficult to provide any interpretations, but that the joint configuration of sales points and subjects can readily be interpreted.

Table 11.7 *Squared Distances between Subjects and Sales Points*

Point	A	B	C	D	E	F	G	H	I	J
1	3.4	2.4	0.7	2.0	0.5	0.7	1.5	1.3	1.9	3.3
2	4.6	5.5	3.0	2.6	2.5	3.4	2.2	2.3	3.7	2.0
3	4.9	5.6	2.9	2.8	2.4	3.2	2.3	2.4	3.9	2.5
4	2.4	1.2	0.3	1.7	0.4	0.2	1.3	1.0	1.1	3.4
5	2.3	0.8	0.9	2.3	1.2	0.7	2.1	1.8	1.3	4.4
6	5.8	4.8	2.1	3.7	1.7	2.1	2.9	2.7	3.8	4.8
7	3.2	1.4	0.9	2.7	1.1	0.7	2.4	2.0	1.8	4.9
8	3.0	1.3	1.0	2.7	1.2	0.8	2.4	2.1	1.7	5.0
9	2.7	1.0	2.1	3.3	2.6	1.8	3.4	3.0	2.0	5.8
10	3.5	1.7	1.0	2.9	1.2	0.8	2.6	2.2	2.0	5.3
11	2.7	1.0	1.3	2.8	1.6	1.0	2.7	2.3	1.6	5.2
12	3.9	3.0	1.0	2.3	0.7	1.0	1.7	1.5	2.3	3.5
13	4.5	3.2	1.2	3.0	1.0	1.1	2.3	2.0	2.7	4.5
14	1.0	0.3	0.1	0.7	0.2	0.0	0.6	0.4	0.3	2.0
15	4.2	5.2	2.8	2.3	2.3	3.2	1.9	2.1	3.4	1.7
16	4.2	2.8	1.0	2.9	0.9	0.9	2.3	2.0	2.5	4.6
17	3.1	3.3	1.3	1.5	0.9	1.5	1.0	1.0	2.1	1.6
18	2.7	1.0	2.1	3.3	2.6	1.8	3.4	3.0	2.0	5.8
19	1.8	2.7	1.4	0.7	1.2	1.8	0.6	0.7	1.5	0.4
20	1.0	1.8	1.0	0.3	0.9	1.3	0.2	0.3	0.8	0.1
21	4.8	3.2	1.3	3.3	1.2	1.2	2.7	2.3	2.9	5.1
22	5.7	4.8	2.0	3.6	1.7	2.1	2.9	2.6	3.8	4.7
23	3.2	1.4	1.2	3.0	1.4	0.9	2.7	2.3	1.9	5.4
24	3.5	4.3	2.2	1.8	1.8	2.6	1.4	1.6	2.8	1.4

Look at three subjects who are in the upper right corner of the first quadrant: they are closest to "Less than 10,000 yen a night." A large cluster of subjects is closest to "Known for excellent food." Another large cluster of subjects is closest to "Scenery." Altogether, it looks as though "Food," "Scenery" and "Outdoor hot spring" seem to be very appealing to many subjects. There is another cluster of people who want to stay with their pets during their hot spring visit, and these people are also equally close to "Sports/exercise facilities," "Resort hotel" and "Beautiful scenery."

Imagine what kind of a joint configuration we must have for a uni-dimensional representation to hold for the data: all the subjects must be at one end of the continuum and the other sales points scattered over the continuum. In particular, the weights of all subjects must be 1 on

Table 11.8 *Rank-2 Approximation*

Point	A	B	C	D	E	F	G	H	I	J	$D^2(i:k)$
1	10	8	3	7	1	2	5	4	5	9	34
2	9	10	6	5	4	7	2	3	8	1	22
3	9	10	8	5	2	7	1	3	6	4	40
4	9	6	2	8	3	1	7	4	5	10	74
5	9	2	3	8	4	1	7	6	5	10	31
6	10	9	2	6	1	3	5	4	7	8	4
7	9	4	2	8	3	1	7	6	5	10	24
8	9	4	2	8	3	1	7	6	5	10	18
9	6	1	4	8	5	2	9	7	3	10	12
10	9	4	2	8	3	1	7	6	5	10	12
11	7	1	3	9	5	2	8	6	4	10	28
12	10	8	2	7	1	3	5	4	6	9	46
13	10	8	3	7	1	2	5	4	6	9	12
14	9	5	2	8	3	1	7	6	4	10	122
15	9	10	6	4	5	7	2	3	8	1	50
16	9	7	3	8	2	1	5	4	6	10	16
17	9	10	4	5	1	6	3	2	8	7	128
18	6	1	4	8	5	2	9	7	3	10	12
19	9	10	6	4	5	8	2	3	7	1	82
20	7	10	8	3	6	9	2	4	5	1	138
21	9	7	3	8	2	1	5	4	6	10	16
22	10	9	2	6	1	3	5	4	5	8	8
23	9	3	2	8	4	1	7	5	7	10	14
24	9	10	6	5	4	7	2	3	8	1	50

A=Day; B=10,000yen; C=Outdoor; D=Family; E=Scenery;
F=Food; G=Sports; H=Resort; I=Recreation; J=Pet

Component 1 and 0 for any other components. This is far from our current example, in which subjects are widely scattered in multi-dimensional space.

Table 11.9 *Rank-2 Approximation of Responses of Subject 6*

Point	A	B	C	D	E	F	G	H	I	J
Data	10	9	1	5	2	3	6	4	7	8
Approximated	10	9	2	6	1	3	5	4	7	8

A=Day; B=10,000yen; C=Outdoor; D=Family; E=Scenery;
F=Food; G=Sports; H=Resort; I=Recreation; J=Pet

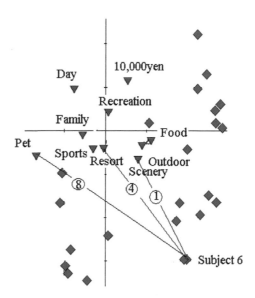

Figure 11.1 *First Two Components and Subject 6*

CHAPTER 12

Successive Categories Data

12.1 Example

This data type has a very strict condition: all the stimuli or objects are judged in terms of a *single set* of ordered categories (e.g., never, sometimes, often, always; strongly disagree, moderately disagree, neutral, moderately agree, strongly agree). Successive categories data can be regarded and analyzed as multiple-choice, which in many cases may be a preferred way of handling the data. The only reason why we identify successive categories data differently from multiple-choice data is because we wish to determine the locations of category boundaries as well as locations of the stimuli.

Table 12.1 is an example in which 3 subjects judged 6 architectural designs (A-F), in terms of three categories (1=poor, 2=good, 3= excellent). The element in the i-th row and j-th column indicates the rated category number of Stimulus j by Subject i.

Table 12.1 *Successive Categories Data*

Subject	A	B	C	D	E	F
1	3	1	2	3	3	2
2	2	1	3	3	3	1
3	1	1	2	2	3	1

12.2 Some Basics

The Thurstonian method of successive categories data (e.g., Bock and Jones, 1968) was adopted as the framework for MUNDA by Nishisato (1980b) and Nishisato and Sheu (1984). First, we assume a uni-dimensional continuum along which successive category boundaries, say

Table 12.2 *Rating of Four Object by Subject 1*

Object	X_1	X_2	X_3	X_4
Response from Subject 1	poor	excellent	good	poor

τ_1, τ_2, ...,τ_m, are located. Each object is rated in terms of $(m + 1)$ categories. Suppose that Subject 1 rated four objects, X_1, X_2, X_3, X_4, as poor, excellent, good, poor, respectively, in terms of the successive category response set [poor, good, excellent] (Table 12.2).

Noting that τ_1, for example, is the boundary between poor and good, we can rearrange the data on a continuum as in Table 12.3. We can

Table 12.3 *Successive Categories Responses from Subject 1*

poor	good	excellent
X_1,X_4	X_2	X_3

now convert the above data to the rank order of category boundaries τ_k and objects X_j on the continuum. Note that X_1 and X_4 occupy the first location, and therefore they share the average rank of rank 1 and rank 2, that is, 1.5. Then, the third position is the first category boundary, thus we give rank 3 to τ_1. Then comes X_2, which receives rank 4. Category boundary τ_2 is ranked 5, and the last position is occupied by X_3, which is thus ranked 6. Table 12.4 shows the summary of the ranked responses. Therefore, the successive categories data can be converted to the ranking of both category boundaries and objects. Once we derive a rank order table, the rest of the analysis can be treated as MUNDA of rank-order data. A major difference of this type of data and rank-order

Table 12.4 *Ranking of Boundaries and Objects*

τ_1	τ_2	X_1	X_2	X_3	X_4
3	5	1.5	4	6	1.5

Table 12.5 *Ten Criminal Acts*

Arson (Ar)	Bribery (Br)
Burglary (Bu)	Forgery (Fo)
Homicide (Ho)	Kidnaping (Ki)
Mugging (Mu)	Raping (Ra)
Receiving stolen goods (Re)	Tax evasion (Ta)

data is that the ranking of category boundaries and objects was derived by using a single unidimensional continuum. In other words, we impose the weak order constraint on the locations of the category boundaries just to generate input rank-order data. Since the weak order constraint is not used in deriving MUNDA weights, there is no guarantee that the order of category boundaries will be maintained in the output. As we will see shortly, however, the category boundaries associated with the first component are likely to be correctly ordered. What can we do if we need more than one component and the category boundaries are not expected to be correctly ordered? We will discuss later what to do when data are obviously multidimensional.

12.3 Seriousness of Criminal Acts

Ten criminal acts (Table 12.5) were rated by 12 subjects. Each subject rated these criminal acts in terms of the following successive categories,

1 = not serious; 2 = moderately serious;
3 = very serious; 4 = extremely serious.

The rating table of 10 criminal acts by 12 subjects is as given in Table 12.6. This table is first converted to the 12x13 table of rank-orders of three category boundaries and ten criminal acts by 12 subjects, which is then transformed to the 12x13 dominance table (Table 12.7). Notice that four successive categories for rating of ten criminal acts have produced many tied ranks, hence tied dominance numbers. This is a typical observation when we analyze successive categories data. Thus successive categories data as handled here are not as informative as data collected by ranking of criminal acts. Notice that the dominance numbers of three category boundaries satisfy the following order for every subject.

$$\tau_1 \leq \tau_2 \leq \tau_3 \tag{12.1}$$

Table 12.6 *Rating of Ten Criminal Acts by Four Successive Categories*

	Ar	Br	Bu	Fo	Ho	Ki	Mu	Ra	Re	Ta
1	4	2	3	2	4	4	3	3	1	1
2	4	2	2	2	4	3	3	4	1	1
3	4	2	2	1	4	4	2	4	1	1
4	3	1	2	2	4	3	3	4	1	1
5	3	1	2	2	4	4	3	3	2	2
6	4	2	2	2	4	3	3	4	1	1
7	4	2	2	2	4	4	3	4	1	1
8	3	1	2	2	4	3	3	4	1	1
9	4	2	3	2	4	4	3	4	2	2
10	4	2	3	2	4	4	3	4	1	2
11	3	1	2	2	4	4	3	3	1	1
12	4	1	3	3	4	4	3	4	2	1

Table 12.7 *Dominance Table of Category Boundaries and Criminal Acts*

	τ_1	τ_2	τ_3	Ar	Br	Bu	Fo	Ho	Ki	Mu	Ra	Re	Ta
1	-8	-2	6	10	-5	2	-5	10	10	2	2	-11	-11
2	-8	0	6	10	-4	-4	-4	10	3	3	10	-11	-11
3	-6	2	4	9	-2	-2	-10	9	9	-2	0	-10	-10
4	-6	0	8	4	-10	-3	-3	11	4	4	11	-10	-10
5	-10	0	8	4	-12	-5	-5	11	11	4	4	-5	-5
6	-8	0	6	10	-4	-4	-4	10	3	3	10	-11	-11
7	-8	0	4	9	-4	-4	-4	9	9	2	9	-11	-11
8	-6	0	8	4	-10	-3	-3	11	4	4	11	-10	-10
9	-12	-2	4	9	-7	1	-7	9	9	1	9	-7	-7
10	-10	-2	4	9	-6	1	-6	9	9	1	9	-12	-6
11	-6	0	8	4	-10	-3	-3	11	11	4	4	-10	-10
12	-10	-4	4	9	-12	0	0	9	9	0	9	-7	-7

This is so because we generated rank orders of category boundaries from a unidimensional continuum. This consistency usually leads to the result that Component 1 yields category boundaries in the correct order. This, however, is not the case for the ensuing components.

Analysis of this dominance table results in the outcome as shown in Table 12.8. Notice that Component 1 accounts for 90 percent of the total information in the data. Thus, there is no need for us to worry about subsequent components. This dominance of Component 1 suggests that the majority of subjects made very similar judgments about

Table 12.8 *Summary Statistics of Component 1*

Eigenvalue	.34
Singular Value	.58
Delta	90.49

Table 12.9 *Weights, Boundaries, Scale Values*

S	weight		τ	Criminal Act	Value
1	1.00	1	-.68	Arson	.63
2	1.02	2	-.06	Bribery	-.59
3	.97	3	.49	Burglary	-.17
4	1.01			Forgery	-.37
5	.95			Homicide	.83
6	1.02			Kidnaping	.63
7	1.03			Mugging	.18
8	1.01			Rape	.68
9	1.00			Stolen Goods	-.80
10	1.02			Tax Evasion	-.76
11	1.00				
12	.98				

the seriousness of those criminal acts. If so, the weights of subjects would be all very close to one. Let us look at the weights for subjects, the values of the three category boundaries and the scale values of the seriousness of the criminal acts (Table 12.9). As expected, weights for subjects are all close to one. Notice the category boundaries and the scale values of the criminal acts. To see the outcome, it is best to draw a uni-dimensional continuum and place these values (Figure 12.1). Since all subjects are bunched together around 1, they are not included in the plot. Imagine that each subject is located to the right-hand side of Homicide, meaning Homicide being the most serious criminal act. The other acts follow it in order, that is, Rape, Arson, Kidnaping, Mugging, Burglary, Forgery, Bribery, Tax evasion and Receiving stolen goods. Subjects seem to agree with this ordering without too many exceptions.

But, what if individual differences are not ignorable? We must then

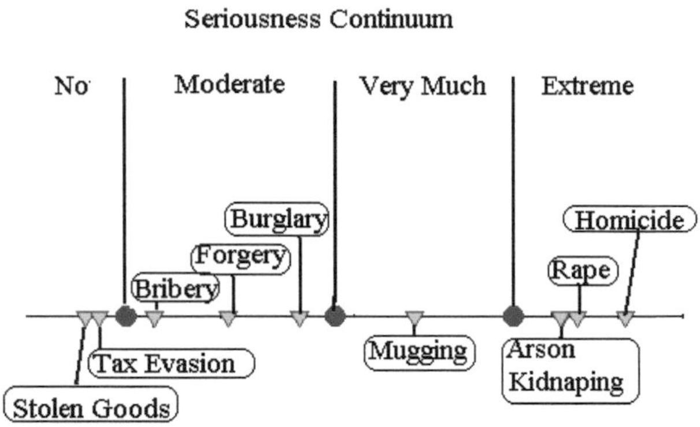

Figure 12.1 *Seriousness of Criminal Acts*

consider multidimensional analysis. The current formulation of MUNDA takes the form of solving an eigenequation, without an order constraint on category boundaries. Thus, Component 2, for instance, is very unlikely to yield correctly ordered category boundaries. What can we do to deal with multidimensional successive categories data?

12.4 Multidimensionality

12.4.1 Multidimensional Decomposition

Conceptually we have no difficulty in seeing such a configuration of successively ordered categories that satisfy an order constraint, not only in unidimensional space (as we have just seen) but also in multi-dimensional space. Consider, for example, two-dimensional space in which ordered categories are located along a line with 45 degree angle to the horizontal axis. They obviously satisfy the correct order with respect to the horizontal axis and the vertical axis. We can easily extend the similar consideration to three- and higher-dimensional space. Our question is how to find a multidimensional configuration of both category boundaries and objects in which category boundaries satisfy an order constraint in each dimension. We need a different algebraic formulation.

Instead, Nishisato (1986b) proposed the following empirical approach.

1. Subject the data to MUNDA and identify subjects whose weights are greater than a pre-specified value such as 0.95. These subjects are identified as Group 1.

2. Subject the data without those subjects in Group 1 to MUNDA and identify those subjects whose weights exceed the pre-specified value, that is, 0.95. These subjects now constitute Group 2.

3. Continue the process until the number of the remaining subjects becomes smaller than a pre-specified number such as 10.

4. Once several groups are identified and the corresponding weights for the category boundaries and objects are obtained, apply these weights of, for example, Group 1 to the entire dominance table to calculate weights of all subjects, say y_1, called here as Component 1; similarly calculate weights for all subjects from the weights of category boundaries and objects obtained from Group 2, y_2, called Component 2; and so on.

5. Calculate correlation between y_j and y_k, say r_{jk} and the angle between Component j and Component k by $\cos^{-1} r_{jk}$.

Nishisato (1986b) showed one numerical example in which the angle between Components 1 and 2 was 93 degrees, rather than the ideal value of 90 degrees. His approach is an empirical one, and further refinement is needed.

Odondi (1997) started with Nishisato's idea for multidimensional decomposition, adopted cluster analysis of subjects, using the dominance table, and refined components through Gram-Schmidt orthogonalization to arrive at a set of orthogonal components that satisfy the order constraint on category boundaries of each component. He provided some numerical examples to show how his method works. Interested readers are referred to his work.

An inferential approach to the multidimensional aspect of successive categories data was proposed by Takane (1981), which is another important contribution made by him. Since the approach is different from that of MUNDA, interested readers are referred to his paper.

12.4.2 Rank Conversion without Category Boundaries

We can ask if the ordered category boundaries are essential for the evaluation of data. Nishisato's study (1980b) was motivated by his desire to formulate a procedure in line with Thurstonian method of successive categories, and not by efficient interpretation of data. So, if we are interested just in the analysis of such data as Table 12.1 or Table 12.6, an easier way is to convert the subjects-by-object rating table into a table of rank orders, and treat it as a rank-order table. We know that we typically use a smaller number of successive categories than the number of objects, thus such rank-order data would contain many tied ranks. Whether we like many tied ranks or not, it is a reflection of the nature of rating data, that is, not much distinct information in the data as compared with rank-order data or paired comparison data. To see what it is like, let us consider converting Table 12.6 into rank-order data. Since we are considering seriousness of criminal acts, let us give Rank 1 to the most serious one. If we look at judgments by Subject 1, there are three criminal acts which were classified into Category 4. The three occupy rank positions 1, 2 and 3. Therefore, we give the average rank of 2 to those three criminal acts. Similarly, Subject 1 rated three criminal acts as 3, and the three occupy rank positions 4, 5 and 6. Since the average rank is 5, we give those criminal acts rank-oder 5. Subject 1 classified two criminal acts into category 2, which occupy rank positions 7 and 8, the average of which is 7.5. Thus those two criminal acts receive rank-order 7.5. The last two criminal acts with rating category on 1 will receive rank-order 9.5. The converted rank-order data are as shown in Table 12.10.

Let us see if we can ignore the category boundaries and still obtain essentially the same results about the seriousness of the ten criminal acts. MUNDA of this converted rank-order data yields the summary statistics as shown in Table 12.11, where only the first two components, out of nine, are listed.

Ninety-one percent of total information is accounted for by Component 1, showing again that we would perhaps need only one component.

Normed weights for subjects and projected weights of the criminal acts, associated with Component 1 are listed in Table 12.12. For ease of comparison, the scale values obtained from MUNDA's successive categories data are listed in the parentheses in the same table. We can tell from Table 12.12 that these scale values from the two modes of analysis are very similar. Considering that the second analysis without category boundaries is capable of handling multidimensional, the

Table 12.10 *Ranked Data of Table 12.6*

	Ar	Br	Bu	Fo	Ho	Ki	Mu	Ra	Re	Ta
1	2	7.5	5	7.5	2	2	5	5	9.5	9.5
2	2	7	7	7	2	4.5	4.5	2	9.5	9.5
3	2.5	6	6	9	2.5	2.5	6	2.5	9	9
4	4	9	6.5	6.5	1.5	4	4	1.5	9	9
5	4	10	7.5	7.5	1.5	1.5	4	4	7.5	7.5
6	2	7	7	7	2	4.5	4.5	2	9.5	9.5
7	2.5	7	7	7	2.5	2.5	5	2.5	9.5	9.5
8	4	9	6.5	6.5	1.5	4	4	1.5	9	9
9	2.5	8.5	5.5	8.5	2.5	2.5	5.5	2.5	8.5	8.5
10	2.5	8	5.5	8	2.5	2.5	5.5	2.5	10	8
11	4	9	6.5	6.5	1.5	1.5	4	4	9	9
12	2.5	10	6	6	2.5	2.5	6	2.5	8.5	8.5

Table 12.11 *Distribution of Information*

Comp	1	2
Eigenvalue	.34	.01
Singular Value	.58	.12
Delta	90.7	3.7
Cum.delta	90.7	94.4

sacrifice of category boundaries seems to be worth making. Although Component 2 is not important, accounting for only 3.7 percent, the second analysis allows us to look at a two-dimensional display (Figure 12.2). The graph shows a typical unidimensional configuration of subjects.

12.4.3 Successive Categories Data as Multiple-Choice Data

We can forget about Thurstonian framework of successive categories data, and subject the data to MUNDA's multiple-choice option. In this case, the data are of incidence type, and we no longer have to deal with dominance numbers. This aspect changes the analysis a great deal. In the current case, homicide was rated 4 by all subjects. As dominance data, it

Table 12.12 *Weights for Subjects, Scale Values of Criminal Acts*

Subject 1	.99	Arson	.584	(.63)
2	1.02	Bribery	-.591	(-.59)
3	.96	Burglary	-.185	(-.17)
4	1.02	Forgery	-.387	(-.37)
5	.94	Homicide	.768	(.83)
6	1.02	Kidnaping	.580	(.63)
7	1.02	Mugging	.149	(.18)
8	1.02	Rape	.623	(.68)
9	1.00	Stolen Goods	-.789	(-.80)
10	1.02	Tax Evasion	-.752	(-.76)
11	1.01			
12	.99			

does not cause any problems. If treated as incidence data, however, it will create a serious problem, due to the fact that the variance of homicide is zero, thus necessitating us to remove the variable from the analysis. Differences between handling the same data as incidence data or dominance data have not been systematically investigated, but it is not difficult to tell that the nature of analysis changes, one dealing with data elements and the other with dominance numbers. More specifically, handling of the data as multiple-choice data means that all options are treated as nominal categories, while if the data are treated as dominance data, ordinal information in the form of dominance numbers is embedded in the processed information. We must then expect that the results will be quite different, and there is no reason to expect any comparable results. As for the current data set, there is no way to compare the results from the two modes of analysis, due to the aforementioned problem with variable homicide, that is, zero variance. So, this comparison of treating the same data as incidence or dominance types is left for the future as an important topic for investigation.

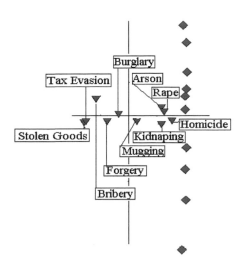

Figure 12.2 *Seriousness of Criminal Acts by Ranking*

PART IV

Beyond the Basics

We have so far discussed most of the basic aspects of MUNDA. However, it is important to remember the very beginning of a motive behind it: we wish to explain a two-way table with a focus on the relation between rows and columns. As discussed so far, MUNDA is said to analyze a two-way table by extracting all possible components, but this does not mean an exhaustive mode of row-column analysis. Indeed, the current formulation of MUNDA does not offer any exhaustive analysis of between-set distances. Recall the problem of discrepancy between the row space and the column space, and consequent problems of joint graphical display. Remember the problem discussed in 5.5.4. Only when MUNDA succeeds in handling both within-set and between-set distances in multidimensional decomposition, we will be able to say that MUNDA offers exhaustive analysis.

The current book has left out many other topics of interest and importance. Some of them, however, will require detailed descriptions to be useful, and others need further work before they can be put into practice.

Basic formulations of MUNDA are available, but as many aliases of MUNDA, discussed in Chapter 3, suggest, a number of new specific applications of MUNDA are constantly emerging, leading to different formulations and bringing out new perspectives for data analysis. Chapter 13 contains brief descriptions of some topics of interest, and Chapter 14 is devoted to future problems of importance.

CHAPTER 13

Further Topics of Interest

13.1 Forced Classification of Dominance Data

Unlike the case of incidence data, dominance data are comparisons of variables, that is, judgments of equality and inequality. Thus, if we want to extend forced classification to dominance data, the criterion must be a pair of variables, rather than a single variable. With a pair, there will be only one proper component, that is, the component that maximizes the difference between variables in the criterion pair.

In discussing an extension of forced classification to dominance data, Nishisato (1984a) has shown that when we multiply a pair of variables by a large enough constant k, the weights of subjects converge to either 1 or -1, depending on whether the subject preferred the first stimulus in the criterion pair or not. That is, given the criterion pair $(X_j, X_{j'})$, the weights of all subjects who prefer X_j to $X_{j'}$ converge to 1, and the weights of those who prefer $X_{j'}$ to X_j to -1. Similarly, if we choose X_j and $X_{j'}$ as the criterion pair in rank order data, the weights of subjects who rank X_j before $X_{j'}$ converge to 1, and the weights of those who rank X_j after $X_{j'}$ to -1. Here, too, there exists only one proper component. These are the asymptotic values of weights as the constant multiplier goes to plus infinity. With this knowledge of the asymptotic weights, we do not have to actually use the forcing agent k, but use those asymptotic weights.

Therefore, forced classification of dominance data can be simplified as that of differential averaging of the dominance table with subjects' weights being either 1 or -1. Forced classification of dominance data, therefore, resembles the choice of the norm employed in Choulakian's taxicab correspondence analysis (Choulakian, 2005).

13.1.1 Forced Classification of Paired Comparisons: Travel Destinations

Let us look at the data in Table 10.4 and carry out forced classification of data using Athens and Honolulu as the criterion pair. Honolulu is a

233

Table 13.1 *Dominance Numbers of Eight Cities with Weights for Subjects*

	y	Athen*	Hono	Istan	Lond	Madr	Paris	Rome	Syd
1	1	-1	-7	3	7	5	-3	1	-5
2	1	5	-7	-3	1	-1	5	5	-5
3	1	-1	-1	-5	1	-5	7	1	3
4	1	1	-1	-7	7	-5	1	5	-1
5	-1	-5	-1	-7	5	3	7	-3	1
6	1	7	-7	5	1	1	1	-3	-5
7	-1	3	5	-3	-3	-3	1	-1	1
8	1	1	-5	-3	5	1	7	1	-7
9	1	3	-5	3	-1	7	-1	1	-7
10	1	5	-5	3	-1	1	-1	3	-5
11	1	5	-5	3	-1	1	-1	3	-5
12	1	5	-7	-1	1	1	-1	7	-5
13	1	7	-7	1	-3	3	-5	5	-1
14	-1	-7	5	-1	3	-3	3	5	-5
15	1	3	1	5	-5	-7	-1	-1	5
16	1	5	3	7	-5	-3	-1	1	-7
17	-1	-7	7	5	-1	-5	-1	-1	3
18	-1	-7	5	-5	3	-3	-1	1	7
19	-1	-7	-5	-3	7	-1	5	1	3
20	1	7	-7	5	-5	-3	1	-1	3

Note: Athen=Athens; Hono=Honolulu, Istan=Istanbul

Lond=London; Madr=Madrid; Syd=Sydney.

Table 13.2 *Weighted Averages of Travel Destinations*

Athens	4.1	Honolulu	-3.8	Istanbul	1.50
London	-1.40	Madrid	0.40	Paris	-0.30
Rome	1.30	Sydney	-2.60		

paradise for outdoor life styles, while Athens is a center of cultural heritage and history. If these are chosen as destinations in the criterion pair, we expect that the results will show an arrangement of the travel destinations on a continuum, which may be called the nature-culture dimension. The results of forced classification can be obtained from the dominance table in Table 10.5 as the weighted averages over twenty subjects, where the weights for subjects who prefer Athens are 1s and those who prefer Honolulu are -1s. Table 13.1 is prepared, using Table 10.5, augmented by weights (1 or -1) for subjects which we can obtain from Table 10.4.

The weighted averages of travel destinations over subjects (Table 13.2) can be used as scale values of the travel destinations, which are plotted on a continuum (Figure 13.1). Athens is followed by Istanbul, Rome, and

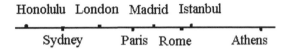

Figure 13.1 *Seriousness of Criminal Acts*

Madrid, while Honolulu at the other end is followed by Sydney. This scale appears to represent the expected dimension.

13.1.2 Forced Classification of Rank-Order Data: Hot Springs

Let us look at the data about sales points of hot springs in Table 11.4. As the criterion pair, let us choose (B: Less than 10,000 yen a night) and (J: Pets allowed to stay), which seem to show a contrast of readiness in spending money. We look at ranking of these two plans by each subject (see Table 11.4) and if the subject ranks B before J (B is preferred to J), the weight for the subject is 1, otherwise -1. The dominance table (Table 11.5) is augmented with weights for the subjects (Table 13.3).

The weighted averages of hot springs over subjects (Table 13.4) were calculated and plotted on a continuum (Figure 13.2). We can see that "Less than 10,000 yen" is followed by excellent food, outdoor hot springs, recreational facilities and beautiful scenery, and at the other end is "Pet allowed to stay," followed by "family baths available," rather than public baths, suggesting preferences for privacy. So, at least we can understand what this arrangement of the hot springs sales points is. As mentioned before, forced classification of dominance data provides only a single proper component. As for conditional components, it looks as though their meanings are not as clear as the case of incidence data.

Table 13.3 *Dominance Table*

		A	B	C	D	E	F	G	H	I	J
1	(1)	-7	-3	-1	-5	9	1	5	7	3	-9
2	(-1)	-7	-9	-3	-5	5	-1	7	3	1	9
3	(-1)	-5	-9	3	-7	7	-1	5	1	-3	9
4	(1)	-7	3	7	9	1	5	-3	-1	-5	-9
5	(1)	-1	9	5	-3	1	7	3	-1	-5	-9
6	(-1)	-9	-7	9	1	7	5	-1	3	-3	-5
7	(1)	-7	7	9	1	3	5	-5	-1	-3	-9
8	(1)	-3	1	5	-7	3	9	-5	-1	7	-9
9	(1)	1	9	-1	-3	5	7	-7	-5	3	-9
10	(1)	-5	5	9	-7	3	7	1	-3	-1	-9
11	(1)	1	5	-1	-7	7	9	-5	-3	3	-9
12	(1)	-7	-5	-1	1	3	9	7	5	-3	-9
13	(1)	-5	-7	3	-3	7	9	-1	5	1	-9
14	(1)	-1	3	-7	1	5	9	7	-9	-3	-5
15	(-1)	-7	-9	-1	7	3	1	-5	5	-3	9
16	(1)	-5	-3	9	-9	7	5	-1	3	1	-7
17	(-1)	-7	-5	1	3	5	7	-9	-1	-3	9
18	(1)	1	9	-1	-3	5	7	-7	-5	3	-9
19	(-1)	-3	-1	-5	5	3	1	9	-7	-9	7
20	(-1)	9	-9	-1	-7	3	1	-5	5	-3	7
21	(1)	-9	1	9	-5	7	5	-1	3	-3	-7
22	(-1)	-9	-7	9	1	7	5	3	-1	-3	-5
23	(1)	-7	9	7	-5	5	3	-3	-1	1	-9
24	(-1)	-9	-3	-1	7	3	1	5	-5	-7	9

A=Day; B=10,000yen; C=Outdoor; D=Family; E=Scenery;
F=Food; G=Sports; H=Resort; I=Recreation; J=Pet

Table 13.4 *Weighted Averages of Hot Springs*

Day	-.58	10,000yen	4.25	Outdoor	1.67
Family	-2.08	Scenery	1.17	Food	3.25
Sports	-.92	Resort	-.42	Recreation	1.29
Pet	-7.33				

What does it mean to partial out the effects of two objects from dominance data? Probably nothing worthwhile.

13.2 Order Constraints on Ordered Categories

In quantification, one popular notion is to impose an order constraints on categories (e.g., Bradley, Katti and Coons, 1962; Inukai, 1972; Nishisato

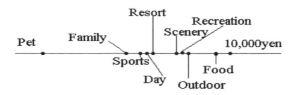

Figure 13.2 *Seriousness of Criminal Acts*

and Inukai, 1972; Williams and Grizzle, 1972; Nishisato and Arri, 1975; Tanaka and Asano, 1978; Tanaka, Asano and Kodake, 1978; Tanaka, Asano and Kubota, 1978; Goodman, 1979, 1981a, b, 1985b; Tanaka and Morikawa, 1980; Tanaka and Kodake, 1980; Kodake, 1980; Clogg, 1982; Agresti, 1983, 1987; Agresti and Kezouh, 1983; Schriever, 1983; Tsujitani, 1984, 1985, 1986, 1987, 1988a, b; Tsujitani and Koch, 1988; Ritov and Gilula, 1993; Beh, 1998, 2001a; D'Ambra, Lombardo and Amenta, 2002; D'Ambra, Beh and Amenta, 2005).

However, when we consider MUNDA and its nature, we wonder if it is useful or even meaningful to impose an order constraint on categories. There are many ordered categories used in collecting data, such as responses to never, rarely, sometimes, often and always. However, is it crucial for us to assign ordered values to them? For instance, -2 for never, -1 for rarely, 0 for sometimes, 1 for often and 2 for always. Or, can we assign values to them in any way that will optimize the criterion such as eigenvalue or reliability coefficient α?

More unfortunately than fortunately, it is a generally accepted view that if the categories are ordered then the weights given to them must also be ordered. Why is the view so popular? It is probably because the mind of many people is still set for unidimensional analysis. Frankly speaking, it is a silly and harmful belief.

Once we leave the realm of unidimensional analysis, it is a free world of scaling in which order constraints may even become detrimental, yes detrimental, to exhaustive analysis of data. Consider the body strength (weak, normal and super strong) as a function of age (infant, teenager, middle-aged and senior). Then, to capture a concave relation between them, we would assign such weights (these are arbitrarily chosen to illustrate the situation) to the categories of body strength as weak= -2.0, normal=0.5, super strong=2.7 (ordered), and those of age as infant= -2.1, teenager=3.1, middle-aged=1.6 and senior=-3.1 (unordered). Then, we know that the body strength is very weak when the subject is an infant or a senior and the strongest when the subject is a teenager or middle-aged. This kind of concave relation between the two variables can never be captured if we were to impose an order constraint on each set of those ordered categories, that is,

$$w_{infant} \leq w_{teenager} \leq w_{middle} \leq w_{senior}$$
$$w_{weak} \leq w_{normal} \leq w_{super}$$

We can ask now if the order constraint imposed on category boundaries in MUNDA of successive categories data is appropriate, or if it is a hindrance for efficient representation of multidimensional successive categories data.

More generally, we should re-consider the wisdom of order constraints in multidimensional data analysis. For example, consider nonparametric item response theory, in which quantification or estimation is carried out by monotone transformation. But, is it indeed what the researchers want? Would you not allow a free estimation procedure if multidimensional analysis is contemplated? Why should we consider only a monotone item characteristic function? Can it not be a multi-valued function instead?

From MUNDA's point of view, order constraints in general data analysis must be seriously looked at in terms of the phenomenon one wants to capture, or the nature of data analysis. In typical data analysis, ordinal measurement does not necessarily mean that only a monotone transformation is allowed.

Of course, an order constraint can be a legitimate choice. For example, consider a linear programming problem, in which the profit of a company needs to be maximized over the expenditures on advertising, expansions of the factory, purchases of new machines, employees' health care benefits, holiday benefits, company's pension contribution, transportation and housing subsidies and profit margins of product sales under the budget constraint (e.g., the budget $\leq$100,000,000 dollars). This constraint that the budget is limited is a real issue, and the constraint must be imposed. One can further introduce such legitimate constraints that, for instance, the advertising cost cannot exceed the transportation subsidy and the latter cannot exceed the housing subsidy.

But in most situations where multidimensional analysis is required, the use of order constraints must be carefully examined and should not be used just because response categories are ordered.

13.3 Stability, Robustness and Missing Responses

MUNDA is a descriptive procedure, and as such our constant concern is with the stability of results of our analysis. Although the stability issue should be considered in conjunction with the sampling procedure for data collection, we often must deal with data as given without much knowledge about the response processes behind generating data. Thus, the stability issue is handled here, not as a way to generalize the results to the population, but more for the sake of anticipating if we can obtain similar results when data are collected in a fashion similar to the original data. Resampling methods such as Jack-knifing and boot-strapping must be evaluated in the above context.

Under some assumptions or models, sensitivity analysis offers added advantage in this regard over those resampling procedures. There are many papers on sensitivity analysis, but we will limit our discussion only to those related to MUNDA.

Tanaka (1978) worked out asymptotic theories of optimal scaling (MUNDA) for investigating sensitivity issues, and there are a large number of contributions to this area by his group as mentioned in Chapter 3, for instance, Tanaka (1983, 1984a, b, 1992), Tanaka and Tarumi (1985, 1988a, b, c), Tarumi (1986), Tarumi and Tanaka (1986) and Huh (1989). Furthermore, Nakayama, Naito and Fujikoshi (1998) considered the stability problem together with analysis using the Hellinger distance as an alternative for stable analysis.

There are a few other topics relevant to sensitivity analysis such as influential observations, perturbation theories, robust methods and missing responses. Out of these, missing responses are what most researchers are faced with, and in the context of MUNDA, missing responses are exceptionally difficult. For, the question is not what values should be imputed for the missing cells, but rather which option, for example, out of five options might have been chosen. Passive imputations, hot-deck imputations, cold-deck imputations, imputations based on the principle of complete ignorance and creating extra options for missing responses are available (Nishisato, 1994). For detailed discussion on procedures for missing responses in MUNDA, refer, for example, to Gleason and Staelin (1975), Escofier and Le Roux, (1976); Nishisato (1980a, 1994), Meulman (1982, 1984), Greenacre (1984), van der Heijden and Escofier

(1989), Gifi (1990), Little and Rubin (1990), Ahn (1991), van Buuren and van Rijckevorsel (1992), Nishisato and Ahn (1995), and Yamada (1996).

As for robust methods, Nishisato (1984b) proposed the "method of reciprocal medians" and the "method of trimmed averages." Although his method of reciprocal medians had a convergence problem, Gabriel and Odoroff (personal communication, 1984a) said that the problem had been solved by their method of median polish. They proposed several robust methods (Gabriel and Odoroff, 1984b) based on weighted and iteratively reweighted least squares methods and the median polish. Nishisato (1986b) also discussed generalized forced classification to suppress outlier responses, and Gibson (1993) examined applications of his procedure to suppress outlier responses. Sachs (1994) used Tukey's biweight to arrive at robust dual scaling. See also studies by Kim (1992) and Nishisato (1994) for different measures of influences or stability of quantification, and another robust method by Choi and Huh (1999).

Stability of results, robust methods, methods for handling missing responses are geared towards correctness of results no matter how one may define "correctness." There is one paper which addresses this correctness issue directly. Adachi (2004a) presented a through investigation into correct classification rates in quantification of multiple choice data. Adachi (2004b) also investigated how nonlinear relations between variables are correctly represented. We have another tool to interpret results correctly, that is, an intelligent clustering technique based on dual scaling by Mucha (2002).

All of these topics discussed in this section remain always as problems that researchers must face, and we need more studies on these topics.

13.4 Multiway Data

Nishisato and Lawrence (1989) presented probably the most general expression for multiway data structure,

$$\mathbf{F} = (\sum \mathbf{P}_i)\mathbf{F}(\sum \mathbf{P_j}) \qquad (13.1)$$

where $\mathbf{F}$ is an $n \times m$ matrix of categorical data, $\mathbf{P}_i$ and $\mathbf{P}_j$ are projection operators to capture row structure and column structure, respectively, such that

$$\sum \mathbf{P}_i = \mathbf{I}_n, \sum \mathbf{P}_j = \mathbf{I}_m \qquad (13.2)$$

It seems that the framework for multiway data matrices cannot be more general than the above. There are a number of papers on multiway data matrices which can be regarded as its special cases. For example, the following studies are similar to the above general framework, although they are formulated for certain specific purposes : forced classification of multiway data (Nishisato, 1984a; Nishisato and Baba, 1999), partial correspondence analysis (Yanai, 1988; Yanai and Maeda, 2002), and constrained quantification (Böckenholt and Böckenholt,1990; Takane and Shibayama, 1991; Böckenholt and Takane, 1994). There are also other outstanding studies using different approaches to multi-way analysis such as Iwatsubo (1974), Yoshizawa (1975, 1976, 1977), Friendly (1994), and Kroonenberg (2001).

Since the topic is related to many types of data analysis, the best source of work is provided by a book by Coppi and Bolasco (1989), which contains many papers presented at the International Meeting on the Analysis of Multiway Data Matrices, held in Rome in 1988.

13.5 Contingency Tables and Multiple-Choice Data

When we have two multiple-choice data, we can derive an important relationship between (a) the contingency table of options of one item (rows) by the options of the other item (columns) and (b) the subjects (rows) by options of the two items (columns) table of (1,0) response patterns. This relation has a direct implication to the CGS scaling (Carroll, Green and Schaffer, 1986, 1987, 1989), or more precisely, to show that the CGS scaling has not solved any problems of space discrepancy. The Carroll-Green-Schaffer papers were published in the *Journal of Marketing Research*. Greenacre's criticisms (Greenacre, 1989), however, did not appear well accepted by those followers of the papers by three eminent scholars. Let us now see a key study relevant to this point.

Nishisato (1980a, pp.84-87) used the following numerical example, and showed an important relation between the quantification of the contingency table and that of multiple-choice data. Responses of 13 subjects to the two multiple-choice questions:

Q1: Do you smoke? yes, no

Q2. Do you prefer coffee to tea? yes, not always, no

resulted in the following three tables, a contingency table (C), a (1,0) response-pattern table (P), and a condensed response-pattern table (D). These tables are equivalent since any one of them can be constructed from any one of the other two. Notice that the two rows of the contingency table are the first two columns of the other tables and the three columns of the contingency table are now the last three columns of

Table 13.5 *Different Data Representation*

C			P					D				
3	2	1	1	0	1	0	0	3	0	3	0	0
1	2	4	1	0	1	0	0	2	0	0	2	0
			1	0	1	0	0	1	0	0	0	1
			1	0	0	1	0	0	1	1	0	0
			1	0	0	1	0	0	2	0	2	0
			1	0	0	0	1	0	4	0	0	4
			0	1	1	0	0					
			0	1	0	1	0					
			0	1	0	0	1					
			0	1	0	0	1					
			0	1	0	0	1					
			0	1	0	0	1					

the other tables. The ranks of the contingency table is generally different from that of each of the other two tables. When we analyze all the three tables, we obtain exactly the same weights for the options of the two items whether they are located in rows or columns.

There are some differences as well. Let us indicate the eigenvalue (correlation ratio) associated with the three formats C, P and D as ρ_c^2, ρ_m^2 and ρ_{mc}^2, respectively. These are related by

$$\rho_c{}^2 = (2\rho_m{}^2 - 1)^2 = (2\rho_{mc}{}^2 - 1)^2 \qquad (13.3)$$

and the scores for subjects, which are lacking in the contingency table, can be recovered as cell scores falling in row i and column j, v_{ij}, by

$$v_{ij} = \frac{y_i + x_j}{\sqrt{2(\rho_c + 1)}} \qquad (13.4)$$

The above results show clearly that the CGS scaling did not solve any problems due to the discrepancy between row space and column space.

Regarding the ranks of these matrices, we note that the rank of the contingency table is two, of which one is accounted for by the trivial solution. Thus, there is one proper component for the contingency table. As for the multiple-choice data, the rank of each of the two tables is 4, of which one corresponds to the trivial solution. Therefore, there are three proper components.

From the relation between the two eigenvalues and noting that the correlation ratio is non-negative, we obtain

$$\rho_m^2 \geq 0.5 \qquad (13.5)$$

In other words, out of those components from the multiple-choice format, the eigenvalue(s) of the component(s) corresponding to the component(s) from the contingency table is (are) greater than or equal to 0.5. Note also that the eigenvalues of the "extra" components from the multiple-choice format are all less than 0.5. What are those extra components? This is worth investigating, for we do not seem to have a full understanding of even this seemingly simple relation between different representation of the same data.

13.5.1 General Case of Two Variables

It is often revealing to use an extreme case when we want to identify characteristics of certain methods. As such, consider the following example of the contingency table (C) and the corresponding response-pattern table (P) (Table 13.6).

From both formats, we will obtain three proper components. This is

Table 13.6 *Contingency (C) and Response-Pattern (P) Tables*

C				P							
3	0	0	0	1	0	0	0	1	0	0	0
0	2	0	0	1	0	0	0	1	0	0	0
0	0	1	0	1	0	0	0	1	0	0	0
0	0	0	2	0	1	0	0	0	1	0	0
				0	1	0	0	0	1	0	0
				0	0	1	0	0	0	1	0
				0	0	0	1	0	0	0	1
				0	0	0	1	0	0	0	1

a special case in which the four columns of the response-pattern table are repeated, thus reducing the rank of the entire table to four, and not 7. A trivial component reduces the total number of proper components by one in both formats. The two tables yield three eigenvalues each, and those eigenvalues are all equal to 1. The three corresponding values of alpha are also all equal to 1, and the values of δ are all equal to 33.33 percent. Notice that this example presents a special case in which both the contingency-table format and the response-pattern format of two categorical variables converge to the identical information content.

Let us divert a little and look at the comparison between categorical data and continuous data. For continuous data, we expect only one component when the value of alpha is one, and it then follows that the value of δ is 100 percent. For categorical data, however, the situation is quite different as we have just observed, and there are in general several components which are needed to exhaustively account for the data. Does this mean that MUNDA should be further developed in such a way that when the value of alpha is 1 (i.e., all categorical scores are perfectly correlated with the component scores) MUNDA should generate only one component which yields δ of 100? If MUNDA can generate such a single component of 100 percent information, it would be very useful and economical. Currently, we can look at the problem in a different way as a comparison of PCA and MUNDA. When we use Likert-type scores for ordered categories, we can use PCA; when the same data are expressed in the format of response-patterns, we can use MUNDA. The two methods yield identical results only when the number of categories of each variable is two, that is, when the data are binary (Nishisato, 1980a). We need a more efficient quantification method than the current formulation of MUNDA.

13.5.2 Statistic δ

From the above discussion, it may be clear that the statistic δ for the contingency table does not behave as one might anticipate from the knowledge of PCA of continuous variables. As the correlation between rows and columns increases, the variance of MUNDA's eigenvalues tends to decrease, contrary to the case with continuous variables. Nishisato (1996) demonstrates that the variance of δ values of the contingency table increases as the table moves away from a diagonal form to the case of statistical independence. In other words, the variance of δ values tends to increase as the information in the table decreases. Thus, one must exercise caution when one interprets the statistic δ from a contingency table.

What about the corresponding response-pattern table? Nishisato (1996) shows that the number of components is equal to the total number of options minus the number of items (variables) and that the total information is equal to the average number of options minus 1. Thus, it is obvious that even when the data produce a perfect value of α or the eigenvalue, the value of δ may not be what one may expect. It is known that the value of δ for response-pattern multiple-choice data may be quite different from what one may expect from the knowledge of PCA of continuous variables.

13.5.3 Extensions from Two to Many Variables

The above discussion on the case of two categorical variables poses an interesting question on how to extend the same formulation to more than two categorical variables. For a two-way table, the singular value decomposition can be expressed also as a bilinear expansion:

$$f_{ij} = \frac{f_{i.}f_{.j}}{f_t}(1 + \rho_1 y_{1i}x_{1j} + \rho_2 y_{2i}x_{2j} + \rho_3 y_{3i}x_{3j} \cdots + \rho_K y_{Ki}x_{Kj})$$
(13.6)

When we extend this line of thought to the case of three categorical variables, it would be for a 3-dimensional contingency table, for which the same decomposition will take the following form,

$$f_{ijk} = \frac{f_{i..}f_{.j.}f_{..k}}{f_t}(1+\rho_1 y_{1i}x_{1j}z_{1k}+\rho_2 y_{2i}x_{2j}z_{2k}+\cdots+\rho_K y_{Ki}x_{Kj}z_{Kk})$$
(13.7)

In this way, it is possible to extend it to an n-dimensional contingency table.

$$f_{ijk...n} = \frac{f_{i....}f_{.j...}f_{..k..}\cdots f_{....n}}{f_t}(1 + \sum_p \rho_p y_{pi}x_{pj}z_{pk}\cdots u_{pn})$$
(13.8)

The problem, however, is computational. Can we find a simple way to solve an n-way singular value decomposition? Can we interpret the results if obtained? What about so many empty cells?

Historically researchers have taken an easier way as Guttman (1941) did. Namely, arrange the n response pattern matrices in a row into the form of subjects-by-options (categories) of all n variables.

$$\mathbf{F} = [\mathbf{F}_1, \mathbf{F}_2, \mathbf{F}_3, \cdots, \mathbf{F}_n]$$
(13.9)

Quantification of this representation of the data changes the relation between the analysis of a contingency table and its corresponding response-pattern matrix. More abruptly speaking, we move away from the quantification problem associated with the contingency table into a new realm of quantification.

13.6 Permutations of Categories and Scaling

In the social sciences, we often use Likert scoring (Likert, 1932) when response categories are ordered. When the chosen category scores are added over a number of questions for each subject and used as the subject's score, the procedure is referred to as the *method of summated ratings*. For this type of data, one can also use those chosen category scores as item scores and subject the data to principal component analysis. These two approaches are severely limited because they capture only linear relations between items. In contrast, MUNDA engages itself in the joint process of permutation and scaling. In other words, MUNDA sees if any re-ordering of categories is necessary to explain the relation between items as linear as possible (permutation of categories) and at the same time if any adjustments of category intervals are necessary (scaling) to maximize their linear relation. In this process, the data matrix is expanded from the subjects-by-item table to the subjects-by-item categories table. Thus, the space spanned by the columns is expanded from n (items) dimensions for linear relations to m dimensions (the sum of all the options of n items) for linear and nonlinear relations. Thus, the difference $m - n$ is introduced to include nonlinear relations. As we have seen earlier, categorization of continuous variables alone does not guarantee that we can tap into non-linear analysis by MUNDA, but that the number of categories of an item must be greater than two. For, the number of categories (options) of an item minus one is related to the degree of a nonlinear relation we can look for.

In the traditional PCA framework, if we were to look into nonlinear relations, one must increase the dimensionality of data, for example, by augmenting the data table by adding to it columns consisting of cross-products of pairs of items, squares of items, triple products of items and so on. However, in practice, such a procedure would introduce many new non-contributing variables. This provides a stark contrast to MUNDA where categories are scaled in relation to the data at hand, thus wasting no variables in describing the data. Indeed MUNDA handles nonlinearity without any difficulty, no matter what response distribution the multiple-choice data may show. For, nonlinear relations are captured by MUNDA through the combination of permutations of categories and scaling. This is indeed an ingenious device of MUNDA.

CHAPTER 14

Further Perspectives

14.1 Geometry of Multiple-Choice Items

Consider a variable with three categories, for example, a question with three response options. We have three possible response patterns, (1,0,0), (0,1,0) and (0,0,1). These can be considered as coordinates in three-dimensional space (Figure 14.1).

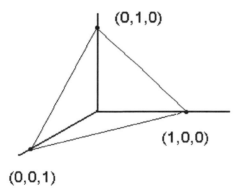

Figure 14.1 *3-Category Variable*

When the data are collected from N subjects, each response falls in only one of these three points, and the final locations of the three coordinates are determined by the chi-square principle: the distance from the origin to the vertex times the frequency is equal for all three vertices. Connecting the three vertices results in a triangle, lying in two-dimensional space. If all the three categories have identical frequencies, the triangle is regular. Similarly, a 4-category variable can be mapped in three-dimensional space, and likewise an n-category variable in $(n\text{-}1)$-dimensional space.

When we consider the cross-product matrix of the $N \times 3$ response pattern data of a 3-category variable, it is a 3×3 contingency table and it is diagonal, which yields one trivial component and two proper components. An important aspect of this is that all non-trivial eigenvalues from a categorical variable are the same and equal to 1. Consider the following three contingency tables, associated with three 3-category variables. We know that two non-trivial eigenvalues are $\rho_1^2 = \rho_2^2 = 1$. One

Table 14.1 *Three Examples of Identical Eigenvalues*

40	0	0	20	0	0	10	0	0
0	40	0	0	80	0	0	40	0
0	0	40	0	0	20	0	0	70

can generalize this to an item of $k + 1$ categories with the property that each item forms $k-$dimensional polyhedron and that each component accounts for the same amount of information, that is, $100/k$ percent.

The above model for a categorical variable provides a stark contrast to PCA of continuous variables, in which each score of a single variable must lie on a straight line going through the origin and the coordinates of the variable. In other words, each variable is represented as an axis. The information of a continuous variable is given by the variance or the length of the vector. In contrast, we regard the area or the volume of a polyhedron as the information associated with the categorical variable in quantification theory.

14.2 A Concept of Correlation

Now that each categorical variable in general is conceptualized as a multidimensional entity, how can we define a measure of correlation? Nishisato (2003b) proposed that the correlation between two categorical variables be defined as the square root of the ratio of the projection of one variable to the other variable. Using our sample data, we can show the projections of the blood pressure and anxiety onto the space of age.

This definition has an astonishing property that such a ratio remains the same over interchanging rows or columns or both of the category-by-category contingency table of two categorical variables. That is, whether the three categories are arranged as 1, 2, 3, or 2, 1, 3, or 3, 2, 1, the correlation of this variable with another variable remains the same. For instance, the following contingency tables are equivalent in the above definition.

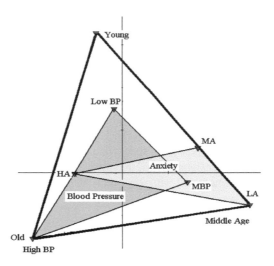

Figure 14.2 *Projection of Blood Pressure and Anxiety onto Age*

Table 14.2 *Four Equivalent Tables*

35	9	7		7	21	12		0	3	41		9	7	35
7	21	12		35	9	7		35	9	7		21	12	7
0	3	41		0	3	41		7	21	12		3	41	0

This definition of correlation, therefore, makes it free from the linear restriction which was implicitly imposed on Pearson's product-moment correlation. The above idea appears conceptually very interesting and appealing, and some attempts were made to operationalize it into a mathematical formula. However, it turned out to lead to some practical difficulties, at least the following ones, and the idea was abandoned for a better alternative.

1. Due to the chi-square distances it is difficult to calculate the area (volume) of the projected polyhedron onto another polyhedron. In the case of a triangle with different frequencies on three vertices, how can we calculate the area of the projection of it onto another triangle with different frequencies on the three vertices?

2. How can we consider projecting, for example, a 4-dimensional polyhedron onto a 5-dimensional polyhedron or vice versa?

3. Greenacre's counter-example. Greenacre (2004) provided an example of a 3×3 contingency table (Table 14.3), which shows the projection of a variable onto the other variable is zero, but the correlation as measured by Cramér's coefficient is 0.47. See the projection of rows onto the column triangle and that of columns onto the row triangle (Figures 14.3 and 14.4). This is the case that the data can be mapped in one-dimensional space, that is, a rank-deficient case, rather than two-dimensional. Thus, in typical analysis, we will not have a triangle, but a line (axis) with row points and column points on it.

14.3 A Statistic Related to Singular Values

It is well known that two categorical variables with numbers of categories m_i and m_j can be fully accounted for by $(p - 1)$ components where p is the smaller number of m_i and m_j and that the full information is contained in the singular values, $\rho_1, \rho_2, ..., \rho_{p-1}$. Thus, to overcome the problems mentioned above, we may consider singular values from which to derive a measure of correlation between two categorical variables.

At this stage, let us look at the square of the item-total (or item-component) correlation again. As we have already seen, it is related to the square of the singular value, or correlation ratio, η^2,

$$\rho^2 = \eta^2 = \frac{\sum_{j=1}^{n} r_{jt}^2}{n} \tag{14.1}$$

In terms of this statistic, the information in our blood-pressure data is distributed as in Table 14.4. As shown earlier, the total contribution of an item in terms of this statistic is equal to the number of categories minus one. Thus, the statistic gives us a straightforward way of dealing

Table 14.3 *Greenacre's Counter-Example, Rows in Column Space*

	1	2	3
1	30	0	0
2	0	15	15
3	10	10	10

with different numbers of options (categories), rather than dealing with triangles and polyhedrons.

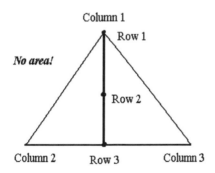

Figure 14.3 *Greenacre's Counter-Example, Columns in Row Space*

As mentioned before, when the sum of the squared item total correlation over all possible dimensions is equal to one (i.e., two categories), one can capture only a linear relation between variables; if the sum is two (i.e., three categories), one can describe linear and

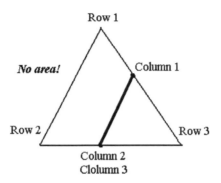

Figure 14.4 *Greenacre's Counter-Example*

quadratic relations between variables. Thus, the sum is related to the power of a polynomial function, up to which nonlinear functional relations between two variables can be captured. Thus, categorizing continuous variables for MUNDA alone does not automatically make it possible to capture any nonlinear relations. Rather, the number of categories of a variable is the key to the question on what nonlinear relations we can capture through quantification.

We now would like to find the distribution of information over a pair of categorical variables. To this end, we will use forced classification (e.g., see Nishisato, 1984a; Nishisato and Gaul, 1990; Nishisato and Baba, 1999). Consider 'Age' as the criterion for forced classification. Then, the distribution of the information as we saw in Table 14.4 changes to Table 14.5. Since the first two so-called proper components of forced classification represents the entire information of item Age, only the relevant portion of the output is listed in the table rather than 12 components.

As defined earlier, these components for which the criterion variable shows perfect correlation are called proper components (dimensions) of forced classification. Look at the entries of these two components of

Table 14.4 *Squared Item-Component Correlation*

Comp	BP	Mig	Age	Anx	Wgt	Hgt	Sum
1	.92	.93	.54	.41	.11	.36	3.27
2	.38	.34	.22	.29	.52	.49	2.25
3	.40	.48	.55	.46	.18	.01	2.07
4	.02	.03	.40	.36	.83	.21	1.84
5	.02	.01	.03	.31	.06	.35	0.78
6	.10	.06	.03	.02	.02	.49	0.72
7	.04	.08	.13	.06	.12	.03	0.45
8	.05	.04	.06	.06	.07	.00	0.28
9	.04	.01	.00	.03	.05	.05	0.19
10	.01	.02	.03	.01	.03	.01	0.10
11	.00	.01	.02	.01	.00	.00	0.04
12	.00	.00	.00	.00	.00	.00	0.00
Sum	2	2	2	2	2	2	12

Table 14.5 *Forced Classification with Age as the Criterion*

Comp	BP	Mig	Age	Anx	Wgt	Hgt	Sum
1	.44	.34	1.00	.56	.12	.07	
2	.35	.07	1.00	.04	.20	.06	
3	.66	.60	.00	.14	.51	.59	
	.	.	.	.	.	.	
	.	.	.	.	.	.	
12	.	.	.	.	.	.	
sum	2.00	2.00	2.00	2.00	2.00	2.00	

'Blood Pressure' and 'Age, that is, 0.44 and 0.35 for BP and 1.00 and 1.00 for Age (criterion). The correlational information between 'Blood Pressure' and 'Age' must be contained in these elements, for the remaining ten components do not contain any information about the criterion variable, 'Age.' Thus we are now ready to consider correlation between categorical variables.

14.4 Correlation for Categorical Variables

14.4.1 A New Measure ν

Let us first study the distribution of information under forced classification. Paying attention only to those proper components, we obtain the following summary of the relevant results, obtained by specifying each item in turn as the criterion, that is, six forced classification outcomes (Table 14.6). Remember that the number of proper components is equal to the number of categories of the criterion variable minus one, which is 2 in the current example.

To see a more general case than this case of uniformly 3-option

Table 14.6 *Squared Item-Total Correlation of Six Forced Classification Analyses*

Item	1	2	3	4	5	6
	1.00	1.00	.34	.23	.20	.33
	1.00	.00	.45	.15	.08	.09
sum		1.00	.79	.38	.28	.42
	1.00	1.00	.34	.27	.17	.33
	0.00	1.00	.07	.37	.32	.07
sum	1.00		.41	.64	.49	.40
	.44	.34	1.00	.56	.12	.07
	.35	.07	1.00	.04	.20	.06
sum	.79	.41		.60	.32	.13
	.23	.53	.45	1.00	.13	.08
	.15	.11	.15	1.00	.07	.00
sum	.38	.64	.60		.20	.08
	.18	.43	.06	.08	1.00	.14
	.10	.06	.26	.12	1.00	.17
sum	.28	.49	.32	.20		.31
	.37	.33	.02	.02	.14	1.00
	.05	.07	.11	.06	.17	1.00
sum	.42	.40	.13	.08	.31	

variables, consider the data in Table 9.4, in which the number of options changes from item to item, namely 2,3,4,5 and 6 for items 1,

Table 14.7 *Five Forced Classification Analyses of Data in Table 9.4*

Item 1	2	3	4	5
1.000	**0.074**	**0.134**	**0.149**	**0.130**
0.004	**1.000**	0.580	0.262	0.311
0.070	**1.000**	0.088	0.212	0.385
sum .074		.668	.474	.696
0.082	0.569	**1.000**	0.547	0.208
0.005	0.090	**1.000**	0.522	0.418
0.047	0.009	**1.000**	0.209	0.289
sum.134	.668		1.278	.915
0.005	0.125	0.698	**1.000**	0.325
0.067	0.100	0.129	**1.000**	0.318
0.012	0.050	0.198	**1.000**	0.177
0.065	0.199	0.253	**1.000**	0.516
sum .149	.474	1.278		1.336
0.003	0.266	0.405	0.505	**1.000**
0.020	0.296	0.330	0.366	**1.000**
0.008	0.090	0.106	0.347	**1.000**
0.035	0.034	0.046	0.082	**1.000**
0.065	0.010	0.028	0.037	**1.000**
sum.130	.696	.915	1.336	

2, 3, 4 and 5, respectively. Using this artificial data set, let us carry out five forced classification analyses with each of items 1, 2, 3, 4 and 5 as the criterion, separately, which yields 1, 2, 3, 4 and 5 proper components, respectively. Let us again list only the proper components and the corresponding values of squared item-component correlations (Table 14.7).

Since some figures are very small, we retained three digits below the decimal point in this table, rather than just two. This was done to verify some numerical equalities, key points for the current discussion. First, we have the following relation,

$$\sum_{k=1}^{P_i} r_{ij(k)}^2 = \sum_{k=1}^{P_j} r_{ij(k)}^2 \qquad (14.2)$$

Table 14.8 *Items 3 and 5 as Criteria and Corresponding Statistics*

Criterion	Item 3		Item 5	
	Item 3	Item 5	Item 3	Item 5
Comp.1	1.000	0.208	0.405	1.000
Comp.2	1.000	0.418	0.330	1.000
Comp.3	1.000	0.289	0.106	1.000
Comp.4			0.046	1.000
Comp.5			0.028	1.000
Sum	3.000	0.915	0.915	5.000

where P_i and P_j are the numbers of proper components when the criteria are Item i and Item j, respectively, and $r^2_{ij(k)}$ is the squared correlation of the non-criterion item for Component k. For example, look at Item 3 as the criterion and the corresponding statistics of Item 5 (i.e., .208+.418+.289=.915), and then Item 5 as the criterion and the corresponding statistics of Item 3 (i.e., .405+.330+.106+.046+.028=.915). Notice that the two sums are identical (Table 14.8).

Interestingly enough, the sum of the squared item-component correlation of a non-criterion item over proper components cannot exceed the smaller number of categories of the two variables minus one, that is,

$$\sum_{k=1}^{P_i} r^2_{ij(k)} = \sum_{k=1}^{P_j} r^2_{ij(k)} \leq (p-1) \tag{14.3}$$

where the summation is over proper components, when Item i is the criterion for forced classification.

Since the two categorical variables can also be represented as an $m_i \times m_j$ contingency table, we can look at the sum of the correlation ratios (eigenvalues) of the contingency table, which provides another important relation:

$$\sum_{k=1}^{P_i} r^2_{ij(k)} = \sum_{k=1}^{P_j} r^2_{ij(k)} = \sum \rho^2_k \tag{14.4}$$

where ρ^2_k is the k-th non-trivial eigenvalue of the contingency table.

Based on these relations (14.3) and (14.4), Nishisato (2005a) proposed the following measure of correlation between two categorical variables:

$$\nu_{ij} = \sqrt{\frac{\sum_{j=1}^{P_i} r_{ij(k)}^2}{p-1}} \tag{14.5}$$

where the summation is over proper components. Since this coefficient reflects the sum of correlations of a number of linearly and nonlinearly transformed components of categorical variables, it is a composite measure of complex relations. Remember that categorical variables were transformed for each component so as to maximize the average of the product-moment correlations.

If we apply the above formula to the blood pressure example, we obtain the matrix of nonlinear correlations as shown in Table 14.9. We

Table 14.9 *Nonlinear Correlation of Blood Pressure Data*

	1	2	3	4	5	6
1	1.00	.71	.63	.44	.37	.46
2	.71	1.00	.45	.57	.50	.45
3	.63	.45	1.00	.55	.40	.26
4	.44	.57	.55	1.00	.32	.20
5	.37	.50	.40	.32	1.00	.39
6	.46	.45	.26	.20	.39	1.00

can also calculate the matrix of nonlinear correlations for our data of Table 9.4 as shown in the table. This measure of correlation proposed is bounded between 0 and 1. This is indicative of the fact that it includes not only linear but also nonlinear relations between variables.

Table 14.10 *Nonlinear Correlation of Nishisato-Baba's Data*

	1	2	3	4	5
1	1.00	.27	.37	.39	.36
2	.27	1.00	.58	.49	.59
3	.37	.58	1.00	.65	.55
4	.39	.49	.65	1.00	.58
5	.36	.59	.55	.58	1.00

14.4.2 Cramér's Coefficient V

Cramér (1946) proposed a measure of association between categorical variables, now known as Cramér's coefficient V. This coefficient is given by

$$V = \sqrt{\frac{\chi^2}{f_t(p-1)}} \qquad (14.6)$$

where

$$\chi^2 = \frac{\sum_i^{m_i} \sum_j^{m_j} \left(f_{ij} - \frac{f_{i.}f_{.j}}{f_t}\right)^2}{\frac{f_{i.}f_{.j}}{f_t}} \qquad (14.7)$$

f_t is the total frequency in the contingency table and p is the smaller value of m_i and m_j, the numbers of categories of variables i and j, respectively.

Nishisato (2005a) discovered that his coefficient ν is mathematically identical to Cramér's V. This equality can be explained as follows. Let us indicate the $N{\times}m_i$ (1,0) response-pattern matrix by $\mathbf{F_i}$ and the $N{\times}m_j$ (1,0) response-pattern matrix by $\mathbf{F_j}$. When we subject the contingency table $\mathbf{F_i'F_j}$ to quantification, we obtain the correlation ratios (squared singular values),

$$1 \geq \rho_1^2 \geq \rho_2^2 \geq ... \geq \rho_{p-1}^2 \qquad (14.8)$$

where the first correlation ratio of 1 is the trivial solution, and p is the smaller value of m_i and m_j. Then, there exists the relation that

$$\sum_{k=1}^{p-1} \rho_k^2 = \sum_{j=1}^{P_i} r_{ij(k)}^2 = \frac{\chi^2}{f_t} \qquad (14.9)$$

From these relations, it is obvious that Nishisato's coefficient ν is equal to Cramér's V, that is,

$$\nu_{ij} = \sqrt{\frac{\sum_{j=1}^{P_i} r_{ij(k)}^2}{p-1}} = \sqrt{\frac{\chi^2}{f_t(p-1)}} = V_{ij} \qquad (14.10)$$

It is interesting to note that Cramér's coefficient has not been one of the most popular measures of association in the social sciences, but that the above identity seems to signal its comeback as a potentially useful and popular measure for analysis of categorical variables.

14.4.3 Tchuproff's Coefficient T

Tchuproff (1925) proposed a measure of association of categorical variables, T, given by

$$T = \sqrt{\frac{\chi^2}{f_t\sqrt{(m_i - 1)(m_j - 1)}}} \tag{14.11}$$

Thus, when $m_i = m_j$, T is equal to Cramér's V. Notice, however, that when m_i is not equal to m_j T may not attain the maximal value of 1, but a maximum smaller than 1, more precisely, if m_i is less than m_j, then

$$T_{max} = \sqrt{\frac{m_i - 1}{m_j - 1}} \tag{14.12}$$

T also captures nonlinear association between categorical variables. But, considering that the maximal value of T depends on the balance between m_i and m_j, it is difficult to imagine where T can be more legitimately applied than V or ν for data analysis.

Now that ν derived from MUNDA is identical to Cramér's V, our question is how to effectively use the statistic. Can we decompose a matrix of ν to arrive at meaningful results? Can we use ν to understand data better than ordinary decomposition by MUNDA? How can we use ν in multidimensional data analysis?

Saporta (1976) first proposed 'partial Tchuproff coefficient' for discriminant analysis of nominal data. For three variables, partial Tchuproff coefficient between variables 1 and 3, given variable 2 is

$$T_{13|2} = \frac{T_{13} - T_{12}T_{23}}{\sqrt{(1 - T_{12}^2)(1 - T_{23}^2)}} \tag{14.13}$$

Then, partial coefficients of higher order are calculated recursively from the above equation, for instance,

$$T_{14|23} = \frac{T_{14|2} - T_{13|2}T_{43|2}}{\sqrt{(1 - T_{13|2}^2)(1 - T_{43|2}^2)}} \tag{14.14}$$

Saporta also defined multiple Tchuproff coefficient,

$$T_{1.23\cdots}^2 = 1 - T_{1|23\cdots} = (1 - T_{12}^2)(1 - T_{13|2}^2)(1 - T_{14|23}^2)\cdots \tag{14.15}$$

Saporta's method of discriminant analysis using these coefficients offers one way of looking at the structure of data, captured by Tchuproff's coefficient. The same idea can readily be extended to Cramér's coefficient.

Let us pause a moment and summarize several interesting properties of squared item-total correlation, which surfaced prominently again in defining Nishisato's ν as well as other statistics such as Cronbach's reliability coefficient α.

14.5 Properties of Squared Item-Total Correlation

When n standardized continuous variables are subjected to PCA, we have the following relations:

$$\sum_{j=1}^{n} r_{jt(k)}^2 = \lambda_k = \text{eigenvalue of component k} \qquad (14.16)$$

$$\sum_{k=1}^{n} r_{jt(k)}^2 = 1 \text{ for any item } j \qquad (14.17)$$

When multiple-choice items with m_j categories are subjected to MUNDA, we have

$$\sum_{j=1}^{n} r_{jt(k)}^2 = n\rho_k^2 \text{ for dimension } k \qquad (14.18)$$

$$\sum_{k=1}^{m-n} r_{jt(k)}^2 = m_j - 1 \text{ for any item } j \qquad (14.19)$$

$$\sum_{j=1}^{n} \sum_{k=1}^{m-n} r_{jt(k)}^2 = n \sum_{k=1}^{m-n} \rho_k^2 = m - n \qquad (14.20)$$

As discussed earlier, the internal consistency reliability, often called 'Cronbach's alpha' (Cronbach, 1951), indicated by α, can be expressed in terms of the squared item-total correlation,

$$\alpha = \frac{n-1}{n} \frac{(\sum_{j=1}^{n} r_{jt(k)^2})}{(\sum_{j=1}^{n} r_{jt(k)}^2 - 1)} \qquad (14.21)$$

Thus, α attains its maximum of 1 when all n items are perfectly correlated with the total score. It is clear that α becomes negative when the sum of squared item-component correlations becomes less than 1. Nishisato (1980a) further indicated that α becomes negative when η^2 is less than $\frac{1}{n}$, which is the average correlation ratio of all possible correlation ratios associated with multiple-choice data.

14.6 Decomposition of Nonlinear Correlation

MUNDA of the six-item blood pressure questionnaire data yields 12 components and associated matrices of inter-item correlation. This means that there are 12 orthogonal mixtures of linear and nonlinear transformations to describe the relation between each pair of variables. Since those 12 correlation coefficients reflect all the relations between two variables, we can logically conjecture that it must be related to the corresponding nonlinear correlation V or ν, which we have just seen to contain all the information about the relation between the two variables. In this regard, V is an upper bound of correlation between two categorical variables.

Looking at the situation differently, we can say that 12 principal axes were used to project each variable and each variable was exhaustively analyzed in 12-dimensional space. If we are to use a geometric representation, six triangles are floating in 6 orthogonal planes and each variable is projected onto six triangles in an exhaustive way. No matter what, 12 coefficients of correlation from MUNDA must contain the entire information about the relation between variables and the same is true for Cramér's coefficient. The difference between a single coefficient of nonlinear correlation ν and the 12 corresponding coefficients of correlation is that the latter 12 coefficients were obtained from the entire set of variables, and that the former from only two variables under consideration. How can we find the relation between a set of 12 correlation coefficients and the corresponding single correlation coefficient ν?

Let us look at only the first four components, out of 12. As we see in Table 14.9, the correlation between Blood Pressures (BP) and Migraines (Mig) is 0.71. The four correlations between BP and Mig in Table 14.11 are 0.99, 0.06, 0.00 and 0.77. By looking at only the first four correlation coefficients from MUNDA, we can already guess that establishing the relation between 0.71 and a set of MUNDA correlation coefficients is very difficult. As noted above, 0.71 was derived from only two variables BP and Mig, while those 12 coefficients were derived from six variables, meaning each of the 12 statistics is under the influence of all the six variables. Thus, perhaps what we should do is to start looking at nonlinear correlation between BP and Mig and its relation to a smaller set of component-wise correlation matrices, such as those obtained from a subset of the data matrix consisting of variables BP, Mig and Age, for instance. Data on these three 3-category variables result in six component-wise correlation matrices. The comparison between

Table 14.11 *Four Correlation Matrices from Quantification*

Component 1	1	2	3	4	5	6
1	1.00					
2	.99	1.00				
3	.59	.58	.1.00			
4	.47	.51	.67	1.00		
5	.43	.39	.08	-.33	1.00	
6	.56	.57	.13	.19	.20	1.00

Component 2	1	2	3	4	5	6
1	1.00					
2	.06	1.00				
3	.59	-.31	1.00			
4	.07	.35	.35	1.00		
5	.28	.62	-.01	.19	1.00	
6	.31	.29	.32	.17	.38	1.00

Component 3	1	2	3	4	5	6
1	1.00					
2	.00	1.00				
3	.67	.27	1.00			
4	.26	.54	.17	1.00		
5	-.02	.43	.14	.13	1.00	
6	.22	-.06	.11	.06	-.32	1.00

Component 4	1	2	3	4	5	6
1	1.00					
2	.77	1.00				
3	-.52	-.46	1.00			
4	-.14	-.14	.48	1.00		
5	.17	.22	.50	.32	1.00	
6	.51	.47	-.13	-.11	.42	1.00

one correlation coefficient and six coefficients is easier than the previous comparison between one and 12 correlation coefficients.

Another matter to remember is the fact that each component-wise correlation matrix is standardized, and therefore that each matrix does not reflect the singular value of that component. In contrast, as we saw in the derivation of ν, this coefficient is composed of values scaled by singular values. Recall that the numerator of Cramér's coefficient is the sum of the eigenvalues of the options-by-options contingency table.

When we look at option weights of each component from MUNDA, we can find out how options of a particular component are transformed, that is, linearly or nonlinearly, this again over 12 components. When we look at option weights of each component from MUNDA of the options-by-options contingency table, we can again tell how options of a particular component are transformed. It may be the case, therefore, that comparisons of option weights from two types of analysis may have some clue for relating the two sets of correlation coefficients.

The rank of our matrix of nonlinear correlation coefficients is six, and we have 12 correlation matrices of total rank of 72. Can we see if there exists such a relation as follows?

$$\mathbf{V} = \sqrt{\sum_{j=1}^{m-n} f(\rho_j) \mathbf{R}_j \bigotimes \mathbf{R}_j} \tag{14.22}$$

where the symbol $\bigotimes$ indicates the element-wise multiplication, that is, the Hadamard product, and $f(\rho_k)$ is a function of the singular value. Currently, the above relation has not been worked out.

Saporta (1975) discussed the orthogonal decomposition of the matrix of Cramér's coefficients or Tchuproff's coefficients. Although his idea does not seem to have been widely put into practice, it offers an interesting possibility for multidimensional decomposition of nonlinearly related data. The idea is to decompose a matrix of rank 6 into linear components. Let us look at PCA of the matrix of ν in Table 14.9, that is, the matrix of nonlinear correlations from the blood pressures data. The results are as shown in Table 14.12. The first two components are dominant, and the remaining components are similar in the values of δ except the last component. The projected weights of the six variables on components are as in Table 14.13. If we look at the first two components and plot the items, we see four clusters, (blood pressures, migraines), (age, anxiety), (height) and (weight), as seen in Figure 14.5. This plot does not seem to tell us anything interesting, or worth even paying attention to, other than perhaps indicating that

Table 14.12 *Information about Components*

Component	1	2	3	4	5	6
Eigenvalue	3.27	.91	.65	.54	.46	.17
δ	54.6	15.0	10.9	9.0	7.6	2.9
Cumδ	54.6	69.7	80.5	89.5	97.1	100.

Table 14.13 *Projected Weights (Loadings) of Variables*

Component	1	2	3	4	5	6
Item 1	.84	.01	.35	.17	.29	.25
Item 2	.85	-.04	.07	-.33	.32	-.24
Item 3	.75	.37	.01	.49	-.17	-.17
Item 4	.70	.49	-.09	-.39	-.31	.12
Item 5	.66	-.30	-.67	.10	.09	.08
Item 6	.60	-.66	.25	-.04	-.37	-.02

through nonlinear transformation we succeeded in maximizing the relations among the items. The discrepancy between the ranks of the matrix of Cramér's coefficients, 6, and those of 12 correlation matrices suggests that we are expanding the dimension of the space to accommodate nonlinear relations. Thus, eigenvalue decomposition of the 6x6 matrix of Cramér's coefficients cannot be justified for our pursuit of nonlinear relations.

Without knowing the relation between MUNDA's 12 correlation matrices and Cramér's correlation matrix, it is difficult to tell which correlation matrix for categorical data we should use for subsequent analysis such as structural equation modeling (SEM). Nishisato and Hemsworth (2002) and Hemsworth (2004) suggested the use of MUNDA's correlation matrix associated with Component 1. But, we have already noted that the correlation matrix is dependent on what other variables are included in the data set, the important point, however, being that this correlation matrix is always positive definite or positive semi-definite. Or, should we use the matrix of Cramér's coefficients or Tchuproff's? Without knowing what transformations are involved, the correlation matrix may not offer interesting component structure. Or, should we use the matrix of polychoric correlation? Do we have a satisfactory procedure to incorporate the underlying multivariate normal distribution to calculate polychoric correlation? If there is a good rationale for decomposing the matrix of Cramér's coefficients into orthogonal components, it would offer a possible use of it for SEM. But what are those multidimensional components of Cramér's coefficient? There should definitely be nonlinear components involved in it.

When Saporta (1976) discussed the use of the Tchuproff coefficient for discriminant analysis of nominal data, he defined partial Tchuproff coefficient, to isolate the influence of other variables from the relation between two variables of interest. A close look at his work may have some hint for multidimensional analysis of Tchuproff's coefficient or Cramér's coefficient.

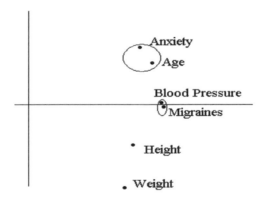

Figure 14.5 *Projection of Blood Pressure and Anxiety onto Age*

A perhaps more direct way to look into analysis of linear and nonlinear relations than the above suggestion is the idea of projection pursuit (Friedman and Tukey, 1974). At the current moment, no one seems to have applied the procedure to categorical data. In a more structured situation than the categorical case, Mizuta (2002) and Mizuta and Hiro (2004) advanced theory of relative projection pursuit to partition data into the normal distribution and the non-normal distribution parts. Thus, if we can identify some basis for projecting data onto the space accountable for by linear relations and the space associated with nonlinear relations, it will advance our knowledge in a more direct way than finding the decomposition of a matrix of Cramér's coefficients into 12 component correlation matrices. This, however, is only an idea and we need further exploration.

An idea similar to the projection pursuit was advanced by the forced classification of dual scaling (Nishisato, 1984a; Nishisato and Gaul, 1990; Nishisato and Baba, 1999), in which proper components and conditional components were isolated. As noted earlier, this was also related to partial correspondence analysis (Yanai, 1988; Yanai and Maeda, 2002). Our object for the current problem is to find a way to project data onto linear space and also onto nonlinear space. If this task can be carried out successfully, it will clarify an interesting relation between linear and nonlinear modes of analysis. Currently MUNDA carries out analysis component by component in which different variables undergo different transformations. Nishisato (1986a) further generalized his procedure into generalized forced classification, and this scheme needs to be further investigated since it has a great possibility as a means of general projection pursuit.

14.7 Interpreting Data in Reduced Dimension

The most common practice in handling multidimensional data is to look at the data structure in reduced space, typically in two or three dimensions, rather than in total multidimensional space. One of the main reasons for this practice is that we can take advantage of graphical display of the distribution of variables, in particular in two-dimensional space. Another reason is that it is difficult to look at all components extracted by such a procedure as factor analysis or MUNDA. The simplicity and interpretability indeed provide us with the practicality of any scientific tools. 'Let us look at only a few major factors' is a ubiquitous scene in the social science research, for it is a widely accepted procedure. However, one should ask if such a practice is indeed valid for understanding of the data. This question has rarely been raised by researchers, which alone should not be taken as the practice being acceptable.

Nishisato (2005c) raised the question if the practice is a valid procedure for data analysis, for in his view the validity of analytical results do not come from either 'simplicity' or 'interpretability' of the procedure. His question is based on two facts:

1. The distance between two variables, viewed in the first two dimensional space, generally increases as we look at it in three-, four- and higher-dimensional space. In other words, by looking at data in reduced space the variables look closer to one another than they actually are.

2. As inferred from the above point, the correlation between two variables looks larger in a smaller-dimensional space than in the full space. In other words, the correlation between two variables viewed in two-dimensional space, for example, looks larger than it actually is in multidimensional space.

This aspect of correlation in multidimensional space can be understood if we remember that in the orthogonal coordinate system, each continuous variable or each nonlinearly transformed categorical variable can be expressed as an axis: Recall that each variable can be regarded as a linear combination of orthogonal components with normalized weights, and that any linear combination (e.g., a composite score) can be expressed as an axis. The product-moment correlation is the cosine of the angle between the two axes. It is easy to imagine that the angle between two axes in multidimensional space cannot become larger (i.e., smaller correlation) in smaller-dimensional space, but tends to become smaller (i.e., larger correlation) in smaller-dimensional space. Thus, we arrive at the second point in the above list.

However, the above point is generally not taken very seriously, but just as an interesting observation. To make the situation very clear, consider that all the inter variable correlation coefficients are close to zero. This is a situation that we need all the components to explain the data. Even in this case, however, if we decide to look at only the first two dimensions, for example, we will suddenly notice substantial amounts of correlation between variables. Is this not a frightening observation, rather than just interesting?

Despite the general view that by looking at data in reduced space we are using less information in data, the above two points suggest that on the contrary we are using too much or inflated information in interpreting data: for the correlation in two-dimensional space, for example, is larger than the 'true' correlation and the distance between two points looks closer than it actually is in the total space. To the best of the author's knowledge, the problem of how much the correlation assessed, for example, in two dimensions, is an over-estimate of the correlation in the total space has not been investigated. The consequences of interpreting data in reduced space are worth investigating and the current practice is likely to be detrimental to the purpose of extracting 'valid' information from data. Considering the practice being widely used and accepted as the procedure for interpreting multidimensional data, its undoing may cast doubts on many conclusions derived from it.

In the social sciences, it is common to interpret outcomes of analysis factor by factor. The procedure is based on looking at variables at both extremes of each factor (axis), and from the contrast of the two sets of such highly loaded variables researchers interpret the axis as representing, for example, the introvert-extravert dimension or factor. This procedure has been popular in the social sciences, and the results often seem to make sense. One of the reasons for many studies on the rotational methods in factor analysis is due to the fact that the factor analytic model is invariant over the rotation of axes. In other words, once we know how to reproduce the off-diagonal elements of the correlation matrix, we can rotate the factor analysis components in any orthogonal way we like without diminishing the capability of reproducing the correlation. Because of this rotational invariance, we naturally developed the idea of axis-interpretation, the practice that attaches a special meaning to each axis.

However, considering serious concerns about interpreting data in reduced space, it seems important to reconsider the traditional 'axis-as-a-factor' interpretation and adopt an alternative procedure based on identifying clusters of variables in multidimensional space. This means that we no longer interpret the results dimension by dimension, but look for clusters of variables in multidimensional space. Researchers should look into the spatial configuration of such clusters. In this way, the choice between oblique axes and orthogonal axes in factor analysis may no longer be meaningful. Using again the above case of an extreme situation with little inter variable correlation, we can say that an axis-interpretation would still lead to a few major factors, while a cluster interpretation based on multidimensional space is likely to lead to the right interpretation that there is no major cluster.

Consequences of interpreting data in reduced space are many, some of which pose serious dangers to the validity of interpretation. This topic, however, is too large to discuss in the current book, and in addition we know too little to discuss it sensibly. Thus, this is another topic left for the future.

14.8 Towards an Absolute Measure of Information

14.8.1 Why an Absolute Measure?

Let us consider the blood pressure data again. All six items have three categories (options) each, and we can represent each variable as a triangle in two dimensions. Thus, in our example we can visualize that six triangles are floating in multidimensional space. Consider two extreme cases, one in which all inter-variable correlation coefficients are one, that is, the case of perfect correlation, and the other in which each triangle occupies two dimensions without an overlap with any other triangles, that is, the case in which all inter-variable correlation coefficients are zero. In the first case, six triangles occupy the same

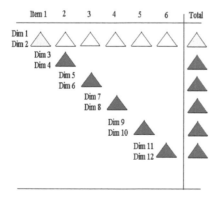

Figure 14.6 *Six Triangles Occupy the Same Space or Distinct Spaces*

two dimensional space, thus we can express this case as in the first row of triangles in Figure 14.6. The total space is also represented by one triangle. In the second case, each triangle occupies two dimensions without any overlap with other triangles. Thus, we can express this case as six triangles occupying in the diagonal locations in Figure 14.6. In this case, the dimensionality of the total space must be 12 since each variable occupies two dimensions. Are these two cases, therefore, not different in terms of the total information they carry? A popular measure of total information is given by the sum of eigenvalues, which is equal to the sum of variances of all the variables. In terms of this definition, the above two cases contain exactly the same amount of information. Does it not sound strange? What is important is that this popular measure of total information is a relative measure, and cannot answer such question as 'Which data set out of two with the same number of variables contains more information?' or 'Which method out of two captures more information?'.

Earlier we demonstrated usefulness of squared item-component correlation as a measure of information conveyed by an item for a given component. Gifi (1990) calls this statistic 'discrimination parameter' in homogeneity analysis. For categorical data, we have seen that the sum of this statistic over all possible components is equal to the number of categories of the variable minus one (Nishisato, 2003a, b, 2005b). So long as we talk about the information conveyed by one variable, this statistic and the variance are useful measures of information.

When we talk about the total information conveyed by a set of variables, however, we would like to advance the idea of taking into consideration the notion of covariation among variables. From this standpoint, we can then say that if all the variables are perfectly correlated to one another, the set contains less information (one triangle in Figure 14.4) than the case in which variables are uncorrelated to one another (six triangles in Figure 14.4). In another way of stating this difference is that if six items are perfectly correlated to one another, we do not need six items, for one item can tell us everything that the six items tell us. These are two extreme cases needed here to make the point that they do not contain the same amount of information, and that for the purpose of comparing two sets of variables in terms of information, we need an 'absolute measure' of information (Nishisato, 2003c).

14.8.2 Union of Sets, Joint Entropy and Covariation

Traditionally, we consider the variation in a data set as information, and define the total information of n variables, $(X_1, X_2, \cdots, X_n)$, by the sum of the variances, that is,

$$\sum_{i=1}^{n} s_i^2 = T(I), \text{say} \tag{14.23}$$

It is known that

$$\sum_{i=1}^{n} s_i^2 = \sum_{i=1}^{n} \lambda_i \tag{14.24}$$

where λ_i are eigenvalues of the $n \times n$ variance-covariance matrix of the n variables.

In terms of set theory, this definition of $T(I)$ is conceptually comparable to the sum of n sets, say

$$A_1 + A_2 + ... + A_n = T_s \tag{14.25}$$

Another concept we can think of is entropy in information theory. When we have a categorical variable assuming values $X_1, X_2, \cdots X_m$, with the corresponding probabilities $P_1, P_2, \cdots P_m$, where $P_i \geq 0$ and $\sum_i^m P_i = 1$. The entropy of the variable is defined as

$$H(X) = -\sum_{i=1}^{m} P_i \log P_i \tag{14.26}$$

When we have n categorical variables, the sum of the entropies can be regarded as a measure of information,

$$H(X_1) + H(X_2) + \cdots + H(X_n) = T_I \tag{14.27}$$

In contrast to the traditional concepts of total information such as $T(I)$, T_s and T_I, we now consider joint information as a measure of total information of a set of variables. If there are three variables, such a measure is

$$\sum_{j=1}^{3} r_{jt(k)}^2 - (r_{12} + r_{13} + r_{23}) + r_{123} = T_r^* \tag{14.28}$$

where

$$r_{123} = \sum_{q=1}^{N} \frac{z_{q1} z_{q2} z_{q3}}{N} \tag{14.29}$$

z_{qj} being the standardized score of Subject q on Item j. Note that we remove overlapping portion of variables from the traditional definition of total information. In set theory, this idea corresponds to considering the union of sets.

$$A_1 \cup A_2 \cup A_3 = T_a^* \tag{14.30}$$

In information theory, this corresponds to the joint entropy,

$$H(X_{123}) = -\sum \sum \sum P_{123} \log P_{123} = T^*_I \tag{14.31}$$

or, in terms of individual entropies,

$$\sum_{i=1}^{3} H(X_i) - H(X_{12}) - H(X_{13}) - H(X_{23}) + H(X_{123}) = T^*{}_I \quad (14.32)$$

where

$$H(X_{ij}) = -\sum\sum P_{ij} \log P_{ij} \quad (14.33)$$

More generally, for n variables, these measures can be expressed as

$$T^*_r = \sum_{j=1}^{n} r^2{}_{jt(k)} - \sum_{i<j}^{n} r_{ij(k)} + \sum_{i<j<m} r_{ijm} - \cdots + (-1)^{n-1} r_{123\cdots n}$$

$$(14.34)$$

where

$$r_{123\cdots p} = \sum_{q=1}^{N} \frac{z_{q1} z_{q2} z_{q3} \cdots z_{qp}}{N} \quad (14.35)$$

z_{qi} is the standardized score of subject q on variable i, and N is the number of subjects. The first term on the left-hand side of the equation is the sum of the discrimination statistics, which is equal to $n\rho^2$. The second term is the sum of inter-variable correlations. The third term corresponds to the sum of triple correlation, defined as the average of the triple products of the corresponding standardized variables, and so on up to the n-tuple correlation, the signs of individual terms alternating. In set theory, the above notion corresponds to the union of sets

$$T^*_s = A_1 \cup A_2 \cup \cdots \cup A_n \quad (14.36)$$

In information theory, it corresponds to the joint entropy,

$$T^*_I = H(X_{123\cdots n}) = -\sum\sum\cdots\sum P_{123\cdots n} \log P_{123\cdots n} \quad (14.37)$$

Consider three variables. If they are perfectly correlated to one another, the sum of the r^2_{jt} is 3, the sum of the inter-variable correlations is 3, and the triple correlation is 1, yielding T^*_r=3-3+1=1. If the three variables are uncorrelated, the first term on the left-hand side is 3, and the second and third terms 0, yielding T^*_r=3-0-0=3. This statistic has not been studies further, and some modifications may be needed to capture what we really want.

Once a valid absolute measure of information is established for a set of variables, it can be used in many situations. For instance, we can use it to see how much more information we can capture by adding two more questions to the questionnaire, which methods out of two can capture more information, how much nonlinear information is contained in data, what are the consequences of data analysis in terms of reduced space, or how much more information MUNDA, for example, captures than PCA when ordered categorical variables are analyzed by both methods. Once we realize that different sets of data contain generally different amounts of information, it is likely that we will find a way to analyze mixtures of different data types such as ranking data and multiple-choice data, typically found in survey instruments, and we will also face the reevaluation of the popular statistic δ, the percentage of total information accounted for by a component, for the base (the total information) of different data sets can be quite different. We should also be able to see how much loss of information standardized MUNDA, as discussed in Chapter 8 under standardized dual scaling, would indeed incur. These are all very important matters for data analysis. Unfortunately, without an absolute measure of information, we cannot provide answers to those questions.

Information retrieval is a key issue for data analysis. Yet, it seems that we do not have a satisfactory measure of information at the current moment. As we deal with massive data, it is always in the mind of researchers that a data reduction technique must be used, which leads to interpreting data in reduced space. We wish we could assess the amount of information distorted by that common practice.

14.9 Final Word

We have looked at multidimensional nonlinear descriptive analysis with illustrative examples of applications. The method described here is in some sense a well-established procedure and widely documented, but as mentioned in this final chapter there are still a number of crucial problems that one must overcome before it is claimed to be a routine multivariate procedure. The fact that one of the most commonly used and widely accepted procedures, that is, multidimensional analysis in terms of reduced space, is problematic must be eye-opening to some researchers. The fact that one of the most commonly used measures of information, that is, the sum of eigenvalues, does not answer a simple question as to which of such two procedures as PCA and MUNDA, applied to the same ordered categorical data, yield more information must be puzzling to some researchers.

It is important to recognize that certain procedures are optimal only for particular situations. In this respect, the current book is most relevant to data sets collected in surveys and observation studies, such as incidence and dominance data types as described in this book. Within these two types of categorical data, however, one would still find many variations of what we have discussed in this book, for which one must find special ways of applying MUNDA. In addition, there are situations in which mixtures of these data types may have to be handled, for which the current book did not discuss any appropriate procedure. This is also one of the problems left for the future.

Since the examination of the relation between rows and columns of a data matrix is MUNDA's task of paramount importance, we cannot be satisfied with the current warning that we should be cautious about comparing distances between rows and columns in symmetric scaling. Instead, we should rigorously pursue the relation between a row and a column by calculating the exact distance (Nishisato and Clavel, 2002), that is, the between-set distance. In other words, our analysis should be directed not only to within-set distances, but also to between-set distances. Only then, we can claim exhaustive analysis of information in data. This is another important task for the future development.

Now that MUNDA can be applied to ordered categorical data, we see a great potential of applying the same procedure to discretized continuous data. In terms of Stevens' classification of measurement, however, continuous data are ratio measurement and discretized continuous data ordinal measurement. Thus, discretization is down-grading of measurement, hence against the object of scaling. In this regard, discretization of continuous data is called desensitization of measurement (Nishisato, 1999). However, Nishisato argues for discretization that it is a step-back in terms of measurement, but that it is a step-up in terms of data analysis. This is so because desensitized data, or discretized continuous data, are now amenable to MUNDA. In his view, the sacrifice of fine information by discretization may be a substantial gain of retrievable information, one of the important gains being nonlinear relations in data.

As for discretization of continuous data, there are a number of publications. Cox (1957) investigated the information loss due to grouping, and studied the effects of grouping continuous variables for two through six intervals. Johari and Sclove (1976) extended the work of Cox to include up to 15-interval measurement for several population distributions. Green and Rao (1970) recommended at least six and preferably eight categories in a configuration recovery study. Jacoby and Matell (1971) considered how many categories would be desirable in test-retest reliability, concurrent validity and predictive validity. Lehmann and Hurlbert (1972) concluded that the use of two- and three-point scales may be appropriate. Stefansky and Kaiser (1973) considered grouping of a normal distribution and retention of information after grouping. Shaw, Huffman and Haviland (1987) extended the work of Stefansky and Kaiser to include more distributions and a new measure of information recovery. There are also many relevant publications in the physical and other sciences, as reviewed by Eouanzoui (2004). Although this topic is very important and relevant to the current book, it was not discussed here because the discussion of the topic can be a book itself. The discretization problem may appear to be simple, but it is a complex one, in particular if we must take into consideration many variables, multidimensionality and non-linear relations. As discussed earlier, the number of categories is relevant to what kinds of nonlinear relations we may consider for the original data, and categorized data require a space of a larger number of dimensions than continuous data. Thus, the problem of discretization is strongly constrained from the practical aspect of data analysis. We should realize that for given sample data, there may not even be an optimal way to discretize them. If not, some compromise must be exercised, leaving doubts about the procedure in researchers' mind. No matter what approach to discretization one may take, it is hoped that it satisfies the basic premise of data analysis, that is, retrieving as much information in data as possible.

When continuous variables are discretized, it is generally believed that the more categories they have the more information they carry. But, from the practical point of view, we have seen that it is not advisable to create many categories. In a totally different situation, some researchers wish to collect more information than one gets from multiple-choice questions, and decide to use open-ended questions such as 'What do you think of the government's tax proposal?' The answers can be whatever the respondents think and are give by statements. We do not know if indeed open-ended questions yield more information than multiple-choice questions. No matter which case it may be, we often face open-ended questions, sometimes mixed with multiple-choice questions. In France, there have been some attempts to develop a technique to handle open-ended questions by correspondence analysis. The technique was rumored to be language-free. Interested readers are referred to Lebart (1982). There is another paper relevant to this problem (Yamanishi and Li, 2002).

Another book is needed to clarify all these problems discussed in this chapter. A viable framework needs to be developed towards comprehensive data analysis, that is, multidimensional nonlinear analysis of continuous and categorical data and such a procedure should not compromise the rich information contained in the input data. The most important basic premise of data analysis is to capture as much information as possible, and this should be done without screening out any substantive information in data. Data analysis needs a common-sense approach.

References

Adachi, K. (2000). A random effect model in metric multidimensional unfolding. *Japanese Journal of Behaviormetrics, 27*, 12-23. (in Japanese).

Adachi, K. (2004a). Correct classification rates in multiple correspondence analysis. *Journal of Japanese Society of Computational Statistics, 17*, 1-20.

Adachi, K. (2004b). Multiple correspondence spline analysis for graphically representing nonlinear relations between variables. *COMSTAT Symposium.* 589-596.

Agresti, A. (1983). A survey of strategies for modeling cross-classification having ordered variable. *Journal of the American Statistical Association, 78*, 184-198.

Agresti, A. (1987). Order-restricted score parameters in association models for contingency tables. *Journal of the American Statistical Association, 82*, 813-828.

Agresti, A. and Kezouh, A. (1983). Association models for multi-dimensional cross-classification of ordinal variable. *Communications in Statistics, A12*, 1261-1276.

Ahn, H. (1991). Effects of missing responses in multiple choice data on dual scaling results. Doctoral thesis, University of Toronto.

Aitchison, J. and Greenacre, M.J. (2002). Biplots in compositional data. *Applied Statistics, 51*, 375-392.

Akiyama, S. (1993). *Suryoka no Graphics - Taido no Tahenryo Kaiseki (Graphics for Quantification: Multivariate Analysis of Attitudes).* Tokyo: Asakura Shoten.

Austin, M.P. (1976). On nonlinear species response models in ordination. *Vegetatio, 33*, 33-41.

Baba, Y. (1986). Graphical analysis of rank data. *Behaviormetrika, 19*, 1-15.

Baker, F.B. (1960). Univac Scientific Computer Program for scaling of psychological inventories by the method of reciprocal averages CPA 22. *Behavioral Science, 5*, 268-269.

Bartlett, M.S. (1947). Multivariate analysis. *Journal of the Royal Statistical Society, Supplement 9*, 176-190.

Beh, E.J. (1997). Simple correspondence analysis of cross-classifications using orthogonal polynomials. *Biomedical Journal, 39*, 589-613.

Beh, E.J. (1998). A comparative study of scores for correspondence analysis with ordered categories. *Biomedical Journal, 40*, 413-429.

Beh, E.J. (1999). Correspondence analysis of ranked data. *Communication is Statistics - Theory and Methods, 28,* 1511-1533.

Beh, E.J. (2001a). Partitioning Pearson's chi-squared statistic for singly ordered two-way contingency tables. *The Australian and New Zealand Journal of Statistics, 43,* 327-333.

Beh, E.J. (2001b). Confidence circles for correspondence analysis using orthogonal polynomials. *Journal of Applied Mathematics and Decision Sciences, 5,* 35-45.

Beh, E.J. (2004). *A Bibliography of the Theory and Applications of Correspondence Analysis.* http:www.uws.edu.au/download.php?file id=7127 &filename= Bibbyyear.pdf&mimetype=application/pdf

Beltrami, E. (1873). Sulle funzioni bilineari (On the bilinear functions). In G. Battagline and E. Fergola (eds.), *Giornale di Mathematiche, 11,* 98-106.

Bennett, J.F. (1956). Determination of the numbers of independent parameters of a score matrix from the examination of rank orders. *Psychometrika, 21,* 383-393.

Bennett, J. F. and Hays, W. L. (1960). Multidimensional unfolding: Determining the dimensionality of ranked preference data. *Psychometrika, 25,* 27-43.

Benzécri, J.P. (1969). Statistical analysis as a tool to make patterns emerge from data. In Watanabe, S. (ed.), *Methodologies of pattern recognition.* New York: Academic Press, 35-74.

Benzécri, J.P. (1979). Sur le calcul des taux d'inertie dans l'analyse d'un questionnaire. *Cahiers de l'Analyse des Données, 4,* 377-378.

Benzécri, J. P. (1982) *Histoire et Préhistoire de l'Analyse des Données* History and prehistory of data analysis. Paris: Dunod.

Benzécri, J. P. (1992) *Correspondence Analysis Handbook.* New York: Marcel Dekker.

Benzécri, J.P. and Benzécri, F. (1980). *L'Analyse des Correspondances: Exposé Elementaire.* Paris: Dunod.

Benzécri, J. P. et al. (1973). *L'analyse des données: II. L'analyse des correspondances.* Paris: Dunod.

Birks, H.J.B., Peglar, S.M. and Austin, H.A. (1994). *An Annotated Bibliography of Canonical Correspondence Analysis and Related Constrained Ordination Methods 1986-1993.* Botanical Institute, University of Bergen.

Birks, H.J.B., Peglar, S.M. and Austin, H.A. (1996). An annotated bibliography of canonical correspondence analysis and related constrained ordination methods. *Abstracta Botanica, 20,* 17-36.

Blasius, J. and Greenacre, M.J. (1998). *Visualization of Categorical Data.* London: Academic Press.

Bock, H.H. (1982). Korrespondenzanalyse zür Strukturerkennung und ihre Verwendung zür Clusteranalyse. *Numerische und Nicht-Numerische Klassifikation zwischen Theorie und Praxis. Proceedings der 5 Fachtagung der Geselschaft für Klassifikation.* Hofgeismar, 36-53.

Bock, R. D. (1956). The selection of judges for preference testing. *Psychometrika, 21,* 349-366.

Bock, R. D. (1960). Methods and applications of optimal scaling. *The University of North Carolina Psychometric Laboratory Research Memorandum,* No. 25.

Bock, R.D., and Jones, L.V. (1968). *Measurement and Prediction of Judgment and Choice.* San Francisco: Holden-Day.

Böckenholt, U. and Böckenholt, I. (1990). Canonical analysis of contingency tables with linear constraints. *Psychometrika, 55,* 633-639.

Böckenholt, U. and Takane, Y. (1994). Linear constraints in correspondence analysis. In Greenacre, M.J. and Blasius, J. (eds.), *Correspondence Analysis in the Social Sciences.* London: Academic Press, 112-127.

Bouroche, J. M. (1977). *Analyse des Données en Marketing.* Paris: Dunod.

Bouroche, J.M., Saporta, G. and Tenenhaus, M. (1975). Generalized canonical analysis of qualitative data. Paper presented at the U.S.-Japan Seminar on Theory, Methods and Applications of Multidimensional Scaling and Related Techniques.

Bradley, R.A., Katti, S. and Coons, I.J. (1962). Optimal scaling of ordered categories. *Psychometrika, 27,* 355-374.

Bradu. D. and Gabriel, K.R. (1978). The biplot as a diagnostic tool for models of two-way tables. *Technometrics, 20,* 47-68. *L'Analyse des Données.* Paris: Que-sais-je?

Browne, M.W. (1976). Personal communication, June 16, 1976.

Burt, C. (1950). The factorial analysis of qualitative data. *The British Journal of Psychology (Statistics Section), 3,* 166-185.

Carleton, T.J. (1984). Residual ordination analysis: a method for exploring vegetation-environmental relationships. *Ecology, 65,* 469-477.

Carlier, A. and Kroonenberg, P.M. (1995). Biplots and decompositions in two-way and three-way correspondence analysis. Technical Report No.01-93. Laboratoire de Statistique et Probabilités, Université Paul Sabatier, Toulouse.

Carlier, A. and Kroonenberg, P.M. (1996). Decompositions and biplots in three-way correspondence analysis. *Psychometrika, 61,* 355-373.

Carliez, F. and Pagés, J. (1976). *Introduction á l'Analyse des Données.* Paris: SMASH.

Carroll, J. D. (1972). Individual differences and multidimensional scaling. In R. N. Shepard, A. K. Romney and S. B. Nerlove (eds.,), *Multidimensional Scaling: Theory and Applications in the Behavioral Sciences,* Volume I. New York: Seminar Press.

Carroll, J.D. (1973). Models and algorithms for multidimensional scaling, conjoint measurement and related techniques. In Green, P.E. and Wind, Y. (eds.), *Multivariate Decisions in Marketing: A Measurement Approach.* New York: Dryden Press, 299-387.

Carroll, J. D., Green, P.E. and Schaffer, C. M. (1986). Interpoint distance comparisons in correspondence analysis. *Journal of Marketing Research, 23,* 271-280.

Carroll, J. D. , Green, P.E. and Schaffer, C. M. (1987). Comparing interpoint distances in correspondence analysis: A clarification. *Journal of Marketing Research, 24*, 445-450.

Carroll, J. D., Green, P.E. and Schaffer, C.M. (1989). Reply to Greenacre's commentary on the Carroll-Green-Schaffer scaling of two-way correspondence analysis solutions. *Journal of Marketing Research, 26*, 366-368.

Celeux, G. and Nakache, J.P. (1994). *Discrimination sur Variables Qualitatives.* Paris: Polytechnica.

Chino, N. (1978). A graphical technique for representing the asymmetric relationships between *N* objects. *Behaviormetrika, 5*, 23-40.

Chino, N. (1990). A generalized inner product model for the analysis of asymmetry. *Behaviormetrika, 27*, 25-46.

Chino, N. (2002). Complex space models for the analysis of asymmetry. In Nishisato, S., Baba, Y., Bozdogan, H. and Kanefuji, K. (eds.), *Measurement and Multivariate Analysis.* Springer: Tokyo, 107-114.

Chino, N. and Shiraiwa, K. (2003). Geometrical structures of some non-distance models for asymmetric MDS. *Behaviormetrika, 20*, 35-47.

Choi, Y.S. and Huh, M.H. (1999). Robust simple correspondence analysis. *Journal of the Korean Statistical Society, 28*, 90-107.

Choulakian, V. (1988). Exploratory analysis of contingency tables by loglinear formulation and generalization of correspondence analysis. *Psychometrika, 53*, 235-250.

Choulakian, V. (2005). Taxicab correspondence analysis. Paper presented at the 70th Annual Meeting of the Psychometric Society, Tilburg.

Cibois, P. (1983). *L'Analyse Factorielle.* Paris: Presses Universitaires de France.

Clausen, S.E. (1998). *Applied Correspondence Analysis: An Introduction.* Sage Publications.

Clogg, C. (1982). Some models for the analysis of association in multiway cross classification having ordered categories. *Journal of the American Statistical Association, 77*, 803-815.

Cochran, W.G. (1954). Some methods for strengthening the common χ^2 test. *Biometrics, 10*, 417-451.

Coombs, C.H. (1950). Psychological scaling without a unit of measurement. *Psychological Review, 57*, 145-158.

Coombs, C.H. (1964). *A Theory of Data.* New York: Wiley.

Coombs, C.H. and Kao, R.C. (1960). On a connection between factor analysis and multidimensional unfolding. *Psychometrika, 25*, 219-231.

Coppi, R. and Bolasco, S. (1989). *Multiway Data Analysis.* Amsterdam: North-Holland.

Cox, C. and Gabriel, K.R. (1982). Some comparisons of biplot display and pencil-and-paper E.D.A.methods. In Luner, R.L. and Siegel, A.F. (eds.). *Modern Data Analysis.*London: Academic Press, 45-82.

Cox, D.R. (1957). Note on grouping. *Journal of the American Statistical*

Association, 52, 543-547.

Craddock, J.M. and Flood, C.R. (1970). The distribution of the χ^2 statistic in small contingency tables. *Applied Statistics, 19*, 173-181.

Cramér, H. (1946). *Mathematical Methods of Statistics*. Princeton: Princeton University Press.

Cronbach, L.J. (1951). Coefficient alpha and the internal structure of tests. *Psychometrika, 16*, 297-334.

Cuadras, C.M. and Fortina, J. (1998). Visualizing categorical data with related metric scaling. In Blasius, J. and Greenacre, M.J. (eds.), *Visualization of Categorical Data*. London: Academic Press, 365-376.

Curtis, J.T. (1959). *The Vegetation of Wisconsin: an Ordination of Plant Communities*. Madison: University of Wisconsin Press.

Curtis, J.T. and McIntosh, R.P. (1951). An upland forest continuum in the prairie-forest border region of Wisconsin. *Ecology, 32*, 476-496.

D'Ambra, L., Beh, E.J. and Amenta, P. (2005). CATANOVA for two-way contingency tables with ordinal variables using orthogonal polynomials. *Communication in Statistics (Theory and Methods)*, 34 (8).

D'Ambra, L. and Lauro, N.C. (1989). Non-symmetrical correspondence analysis for three-way contingency tables. In Coppi, R. and Bolasco, S. (eds.), *Multiway Data Analysis*. Amsterdam: Elsevier, 301-315.

D'Ambra, L. and Lauro, N.C. (1991). Non-symmetrical analysis of three-way contingency tables. In Coppi, R. and Bolasco, S.(eds.), *Multiway Data Analysis*. Amsterdam: North-Holland, 301-315.

D'Ambra, L., Lombardo, R. and Amenta, P. (2002). Singly ordered non-symmetric correspondence analysis. *Act of the XLI Italian Statistics Society Conference*, Milan.

Daudin, J.J. and Trecourt, P. (1980). Analyse factorielle des correspondances et modeles logolinieres:comparaisons de methodes sur un exemple. Revue de Statistique Appliquée, 1.

Davidson, J. (1973). A geometrical analysis of the unfolding model: General solutions. *Psychometrika, 38*, 305-336.

Day, D.A. (1989). Investigating the validity of the English and French versions of the Myers-Briggs Type Indicator. Ph.D. Thesis, University of Toronto.

de Leeuw, J. (1973). *Canonical analysis of categorical data*. Doctoral Thesis, Leiden University.

de Leeuw, J. (1983). On the prehistory of correspondence analysis. *Statistica Neederlandica, 37*, 161-164.

de Leeuw, J. (1984a). The Gifi system of nonlinear multivariate analysis. In Diday, E., Jambu, M., Lebart, L., Pagés, J. and Tomassone, R. (eds.), *Data Analysis and Informatics, III*. Amsterdam: North-Holland, 415-424.

de Leeuw, J. (1984b). *Canonical analysis of categorical data*. Leiden University: DSWO Press.

de Leeuw, J., Heiser, W., Meulman, J. and Critchley, F. (eds.). (1987). *Multidimensional data analysis*. Leiden University: DSWO Press.

de Leeuw, J., Young, F.W. and Takane, Y. (1976). Additive structure in qualitative data: an alternating least squares method with optimal scaling features. *Psychometrika, 41*, 471-503.

Digby, P.G.N. and Gower, J.C. (1981). Ordination between- and within-groups applied to soil classification. In Merriam, D.F.(ed.), *Down-to-Earth Statistics: Solutions Looking for Geological Problems.* Syracuse University Geology, 63-75.

Eckart, C., and Young, G. (1936). The approximation of one matrix by another of lower rank. *Psychometrika, 1*, 211-218.

Edgerton, H.A. and Kolbe, L.E. (1936). The method of minimum variation for the composite criteria. *Psychometrika, 1*, 183-187.

Ellenberg, H. (1948). Unkrautgesellschaften als Mass für den Säuregrad, die Verdichtung und andere Eigenschaften des Ackerbodens. *Berichte über Landtechnik, Kuratorium für Technik und Bauwesen in der Landwirtschaft, 4*, 130-146.

Eouanzoui, K.B. (2004). On desensitizing data from interval to nominal measurement with minimum information loss. Doctoral Thesis, University of Toronto.

Escofier-Cordier, B. (1969). L'analyse factorielle des correspondances. *Bureau Univesitaire de Recherche Operationelle.*

Escofier, B. (1978). Analyse factorielle et distances répondant au principe d'équivalence distributionnelle. *Revue de Statistique Appliquée, 16*, 29-37.

Escofier, B. (2003). *Analyse des Correspondances: Recherches au Coeur de l'Analyse des Données.* Renne: Presses Universitaires de Renne.

Escofier, B. and Le Roux, B. (1976). *Etude de Trois Problèmes de Stabilité en Analyse Factorielle.* Paris: Publication de l'ISUP.

Escofier, B. and Pagés. J. (1988). *Analyse Factorielles Simples et Multiples: Objectives, Méthodes et Interprétation. Paris: Dunod. Cahiers, Série Recherche,* Université de Paris.

Escoufier, Y. and Grorud, A. (1980). Analyse factorielle des matrices carrées non symmetriques. In Diday, E. et al. (eds.), *Data Analysis and Informatics.* Amsterdam: North Holland, 263-276.

Escoufier, Y. and Junca, S. (1986). Discussion of "Some useful extension of the usual correspondence analysis and the usual log-linear models approach in the analysis of contingency tables" by L.A. Goodman. *International Statistical Review, 54*, 279-283.

Evans, G.T. (1970). The analysis of categorizing behavior. *Psychometrika, 35*, 367-392.

Everitt, B.S. (1978). *Graphical Techniques for Multivariate Data.* New York: North-Holland.

Everitt, B.S. (1997). Annotation: correspondence analysis. *Journal of Child Psychology and Psychiatry, 38*, 737-745.

Fasham, M.J.R. (1977). A comparison of nonmetric multidimensional scaling, principal component analysis and reciprocal averaging for the ordination of simulated coenolines and coenoplanes. *Ecology, 54*, 618-622.

Fénelon, J.P. (1981). *Qu'est ce que l'Analyse des Données?* Paris: Lefonen.

Fhanér, S. (1967). Information loss when ρ is applied in non-normal cases. *Scandinavian Journal of Psychology, 8*, 127-131.

Fienberg, S.E. and Meyer, M.M. (1983). Loglinear analysis and categorical data analysis with psychometric and econometric applications. *Journal of Econometrics, 22*, 191-214.

Fisher, R. A. (1940). The precision of discriminant functions. *Annals of Eugenics, 10*, 422-429.

Fisher, R. A. (1948). *Statistical methods for research workers.* London: Oliver and Boyd.

Foucart, T. (1982). *Analyse Factorielle: Programmation sur Micro-Ordinateur.* France: Masson.

Foucart, T. (1985). *Analyse Factorielle: Programmation sur Micro-Ordinateur.* (2nd edition). France: Masson.

Franke, G.R. (1985). Evaluating measures through data quantification: Applying dual scaling to an advertising copytest. *Journal of Business Research, 13*, 61-69.

Friedman, J.H. and Tukey, J.W. (1974). A projection pursuit algorithm for exploratory data analysis. *IEEE Transactions on Computer*, c-23, 9, 881-890.

Friendly, M. (1994). Mosaic displays for multi-way contingency tables. *Journal of the American Statistical Association, 89*, 190-200.

Gabriel, K.R. (1971). The biplot graphical display of matrices with applications to principal component analysis. *Biometrics, 58*, 453-467.

Gabriel, K.R. (1981). Biplot display of multivariate matrices for inspection of data and diagnosis. In Burnett, V. (ed.), *Interpreting Multivariate Data.* New York: Wiley, 147-173.

Gabriel, K.R. (2002). Goodness of fit of biplots and correspondence analysis. *Biometrika, 89*, 423-436.

Gabriel, K.R. and Odoroff, C.L. (personal communication, 1984a)

Gabriel, K.R. and Odoroff, C.L. (1984b). Resistant lower rank approximation of matrices. *Technical Report.* Department of Statistics and Division of Biostatistics, University of Rochester.

Gabriel, K.R. and Odoroff, C.L. (1990). Biplots in biomedical research. *Statistics in Medicine, 9*, 423-436.

Gauch, H.G. (1982). *Multivariate Analysis in Community Ecology.* Cambridge: Cambridge University Press.

Gauch, H.G., Chase, G.G. and Whittaker, R.H. (1974). Ordination of vegetation samples by Gaussian species distribution. *Ecology, 55*, 1382-1390.

Gauch, H.G. and Wentworth, T.R. (1976). Canonical correlation analysis as an ordination technique. *Vegitatio, 33*, 17-22.

Gauch, H.G., Whittaker, R.H. and Singer, S.B. (1981). A comparative study of nonparametric ordinations. *Journal of Ecology, 69*, 135-152.

Gauch, H.G., Whittaker, R.H. and Wentworth, T.R. (1977). A comparative study of reciprocal averaging and other ordination techniques. *Journal of Ecology, 65,* 157-174.

Gibson, L.L. (1993). An investigation of the generalized forced classification procedure and its application to diminishing outlier effects. Master Thesis, University of Toronto.

Gifi, A. (1990). *Nonlinear multivariate analysis.* New York: Wiley.

Gilula, Z. and Haberman, S.J. (1986). Canonical analysis of contingency tables by maximum likelihood. *Journal of the American Statistical Association, 81,* 780-788.

Gilula, Z. and Ritov, Y. (1990). Inferential ordinal correspondence analysis: motivation, derivation and limitations. *International Statistical Review, 58,* 99-108.

Gleason, H.A. (1926). The individual concept of the plant association. *Bulletin, Torrey Botanical Club, 53,* 7-26.

Gleason, T.C. and Staelin, R. (1975). A proposal for handling missing data. *Psychometrika, 40,* 229-252.

Gold, E.M. (1973). Metric unfolding: data requirements for unique solution and clarification of Schönemann's algorithm. *Psychometrika, 38,* 555-569.

Goodall, D.W. and Johnson, R.W. (1982). Nonlinear ordination in several dimensions. A maximum likelihood approach. *Vegetatio, 48,* 197-208.

Goode, F.M. (1957). Interval scale representation of ordered metric scale. Unpublished manuscript, University of Michigan.

Goodman, L.A. (1979). Simple models for the analysis of cross-classifications having ordered categories. *Journal of the American Statistical Association, 74,* 537-552.

Goodman, L.A. (1981a). Association models and canonical correlation in the analysis of cross classifications having ordered categories. *Journal of the American Statistical Association, 76,* 320-334.

Goodman, L.A. (1981b). Association models and the bivariate normal distribution in the analysis of cross-classifications having ordered categories. *Biometrika, 68,* 347-355.

Goodman, L.A. (1985a). Correspondence analysis models, log-linear models, and log-linear analysis models for the analysis of contingency tables. *Bulletins of the International Statistical Institute, 51,* Book 4, 28, 1-14.

Goodman, L.A. (1985b). The analysis of cross-classified data having ordered categories. Association models, correlation models, and asymmetry models for contingency tables with or without missing entries. *The Annals of Statistics, 13,* 10-69.

Goodman, L.A. (1986). Some useful extensions of the usual correspondence analysis and the usual log-linear models approach in the analysis of contingency tables (with discussion). *International Statistical Review, 54,* 243-309.

Goodman, L.A. (1991). Measures, models, and graphical displays in the analysis of cross-classified data (with discussion). *Journal of the American Statistical*

Association, 86, 1085-1138.

Gower, J.C. (1984). Multivariate analysis: ordination, multidimensional scaling and allied topics. In Lloyd, E.H.(ed.), *Handbook of Applicable Mathematics, Vol.VI, Statistics*. Chichester: Wiley, 727-781.

Gower, J.C. (1990). Three-dimensional biplots. *Biometrika, 77*, 773-785.

Gower, J.C. (1992). Generalized biplots. *Biometrika, 79*, 475-493.

Gower, J.C. and Hand, D.J. (1996). *Biplots*. London: Chapman and Hall.

Gower, J.C. and Harding, S. (1988). Nonlinear biplots. *Biometrika, 75*, 445-455.

Green, P.E. and Rao, V. (1970). Rating scales and information recovery - How many scales and response categories to use? *Journal of Marketing, 34*, 33-39.

Greenacre, M.J. (1984). *Theory and applications of correspondence analysis*. London: Academic Press.

Greenacre, M.J. (1988). Correspondence analysis of multivariate categorical data. *Biometrika, 75*, 457-467.

Greenacre, M.J. (1989). The Carroll-Green-Schaffer scaling in correspondence analysis: A theoretical and empirical appraisal. *Journal of Marketing Research, 26*, 358-365.

Greenacre, M.J. (1993a) *Correspondence Analysis in Practice*. London: Academic Press.

Greenacre, M.J. (1993b). Biplots in correspondence analysis. *Journal of Applied Statistics, 20*, 251-269.

Greenacre, M.J. (1994). Multiple and joint correspondence analysis. In Greenacre, M.J. and Blasius, J. (eds.), *Correspondence Analysis in the Social Sciences*. London: Academic Press, 141-161.

Greenacre, M.J. (2000). Correspondence analysis of square asymmetric matrices. *Applied Statistics, 49*, 297-310.

Greenacre, M.J. (2004). Personal communication, March 10, 2004, Dordmund.

Greenacre, M.J. (2005). Thirteen different ways to define correspondence analysis. Paper presented at the 70th Annual Meeting of the Psychometric Society, Tilburg.

Greenacre, M.J. and Blasius, J. (eds.) (1994). *Correspondence Analysis in the Social Sciences*. London: Academic Press.

Greenacre, M.J. and Blasius, J. (eds.) (2006). *Multiple Correspondence Analysis and Related Methods*. Boca Raton: Chapman and Hall/CRC.

Greenacre, M.J. and Browne, M.W. (1986). An efficient alternating least-squares algorithm to perform multidimensional unfolding. *Psychometrika, 51*, 241-250.

Greenacre, M.J. and Torres-Lacomba, A. (1999). A note on the dual scaling of dominance data and its relationship to correspondence analysis. *Working Paper Ref. 430*, Departament d'Economia i Empresa, Universidat Pompeu Fabra, Barcelona, Spain.

Groenen, P.J.F. and Poblome, J. (2002). Constrained correspondence analysis for seriation in archaeology applied to Sagalassos ceramic tableware. In Opitz, O. and Schwaiger, M. (eds.), *Exploratory Data Analysis in Empirical Research*.

Heidelberg: Springer, 90-97.

Groenen, P.J.F. and van de Velden, M. (2004). Inverse correspondence analysis. *Linear Algebra and Its Applications, 388*, 221-238.

Guttman, L. (1941). The quantification of a class of attributes: A theory and method of scale construction. In the Committee on Social Adjustment (ed.), *The prediction of personal; adjustment.* New York: Social Science Research Council, 319-348.

Guttman, L. (1946). An approach for quantifying paired comparisons and rank order. *Annals of Mathematical Statistics, 17,*144-163.

Guttman, L. (1950). Chapters 3-6. In Stouffer, S.A. et al. (eds.), *Measurement and prediction.* Princeton: Princeton University Press.

Guttman, L. (1967). The development of nonmetric space analysis. A letter to Professor John Ross. *Multivariate Behavioral Research, 2,* 71-82.

Han, S.T. and Huh, M.H. (1995). Biplot of ranked data. *Journal of the Korean Statistical Society, 24,* 439-451.

Hand, D.J. (1996). Statistics and the theory of measurement. *Journal of the Royal Statistical Society, Section A, 3,* 445-492.

Hand, D.J. (2004). *Measurement Theory and Practice: The World of Quantification.* London: Arnold.

Hayashi, C. (1950). On the quantification of qualitative data from the mathematico-statistical point of view. *Annals of the Institute of Statistical Mathematics, 2,* 35-47.

Hayashi, C. (1952). On the prediction of phenomena from qualitative data and the quantification of qualitative data from the mathematico-statistical point of view. *Annals of the Institute of Statistical Mathematics, 3,*69-98.

Hayashi, C. (1964). Multidimensional quantification of the data obtained by the method of paired comparison. *Annals of the Institute of Statistical Mathematics, 16,* 231-245.

Hayashi, C. (1967). Note on quantification of data obtained by paired comparison. *Annals of the Institute of Statistical Mathematics, 19,* 363-365.

Hayashi, C. (1974). *Suryoka no Houhou (Methods of Quantification).* Tokyo: Toyo Keizai Sha (in Japanese).

Hayashi, C. (1992). *Suryoka - Riron to Houhou (Quantification: Theory and Methods).* Tokyo: Asakura Shoten. (in Japanese).

Hayashi, C., Higuchi, I. and Komazawa, T. (1970). *Johoshori to Tokeisuri (Information Processing and Statistical Mathematics).* Tokyo: Sangyo Tosho. (in Japanese).

Hayashi, C. and Suzuki, T. (1986). *Shakaichosa to Suryoka (Social Surveys and Quantification.* Tokyo: Iwanami Shoten.

Hays, W. L. and Bennett, J. F. (1961). Multidimensional unfolding: Determining configuration from complete rank order preference data. *Psychometrika, 26,* 221-238.

Heiser, W. J. (1981). *Unfolding Analysis of Proximity Data.* Leiden University: DSWO Press.

Heiser, W.J. (1982). Joint ordination of species and sites: the unfolding technique. In Legendre, P. and Legendre, L.(eds.), *Developments in Numerical Ecology*. Berlin:Springer, 189-221.

Heiser, W.J. (2004). Geometric representation of association between categories. *Psychometrika, 69*, 513-545.

Hemsworth, D. (2004). Modeling using ordinal data: The use of dual scaling in structural equation modeling. Doctoral Thesis, University of Toronto.

Hill, M.O. (1973). Reciprocal averaging: An eigenvector method of ordination. *Journal of Ecology, 61*, 237-249.

Hill, M.O. (1974). Correspondence analysis: a neglected multivariate method. *Journal of the Royal Statistical Society C (Applied Statistics), 23*, 340-354.

Hill, M.O. and Gauch, H.G. (1980). Detrended correspondence analysis: an improved ordination technique. *Vegetatio, 42*, 47-58.

Hirschfeld, H. O. (1935). A connection between correlation and contingency. *Cambridge Philosophical Society Proceedings, 31*, 520-524.

Hoffman, D.L. and Franke, G.R. (1986). Correspondence analysis: Graphical representation of categorical data in marketing research. *Journal of Marketing Research, 23*, 213-217.

Hojo, H. (1994). A new method for multidimensional unfolding. *Behaviormetrika, 21*, 131-147.

Horst, P. (1935). Measuring complex attitudes. *Journal of Social Psychology, 6*, 369-374.

Horst, P. (1936). Obtaining a composite measure from a number of different measures of the same attribute. *Psychometrika, 1*, 53-60.

Horst, P. (1970). Personal communication, September 1970.

Hotelling, H. (1933). Analysis of complex of statistical variables into principal components. *Journal of Educational Psychology, 24*, 417-441, and 498-520.

Huh, M.H. (1989). Local aspects of sensitivity analysis in Hayashi's third method of quantification. *Journal of Japanese Society of Computational Statistics, 2*, 55-63.

Hwang, H. and Takane, Y. (2002). Generalized constrained multiple correspondence analysis. *Psychometrika, 67*, 211-224.

Ihm, P. and Van Groenwoud, H. (1984). Correspondence analysis and Gaussian ordination. *COMPSTAT Lectures 3*, 5-60.

Inukai, Y. (1972). Optimal versus partially optimal scaling of polychotomous items. Master's Thesis, University of Toronto.

Israëls, A. (1987). *Eigenvalue Techniques for Qualitative Data*. Leiden University: DSWO Press.

Iwatsubo, S. (1974). Two classification techniques of 3-way discrete data quantification by means of correlation ratio and three-dimensional correlation coefficient. *Japanese Journal of Behaviormetrics, 2*, 54-65.

Iwatsubo, S. (1975). A review of "Nishisato, S. 'Ouyou Sinrishakudo Kouseiho: Shitsuteki Data no Bunseki to Kaishaku (Applied Psychological Scaling: Analysis and Interpretation of Qualitative Data).

Tokyo: Seishin Shobo, 1975.' *Japanese Journal of Behaviormetrics, 3*, 66.

Iwatsubo, S. (1987). *Suryoka no Kiso (Foundations of Quantification).* Tokyo: Asakura Shoten. (in Japanese)

Jackson, D.N. and Helmes, E. (1979). Basic structure content scaling. *Applied Psychological Measurement, 3*, 313-325.

Jacoby, J. and Matell, M. (1971). Three-point Likert scales are good enough. *Journal of Marketing Research, 8*, 495-500.

Jambu, M. (1989). *Exploration Informatique et Staitistique des Données.* Paris: Dunod.

Jambu, M. and Lebeaux, M.O. (1978). *Classification Automatique pour l'Analyse des Données Logiciels.* Paris: Dunod.

Johari, S. and Sclove, S.L. (1976). Partitioning a distribution. *Communications in Statistics - Theory and Methods, A5*, 133-147.

Johnson, P.O. (1950). The quantification of qualitative data in discriminant analysis. *Journal of the American Statistical Association, 45*, 65-76.

Johnson, R. M. (1963). On a theorem stated by Eckart and Young. *Psychometrika, 28*, 259-263.

Jordan, C. (1874). Mémoire sur les formes bilinieres (Note on bilinear forms). *Journal de Mathématiques Pures et Appliquées, deuxiéme Série, 19*, 35-54.

Jöreskog, K. and Sorbom, D. (1996). *Prelis 2: Users reference guide.* Chicago: Scientific Software International.

Kalantari, B., Lari, I., Rizzi, A. and Simeone, B. (1993). Sharpe bounds for the maximum of the chi-square index in a class of contingency tables with given marginals. *Computational Statistics and Data Analysis, 16*, 19-34.

Keller, J.B. (1962). Factorization of matrices by least squares. *Biometrika, 49*, 239-242.

Kendall, M.G. and Stuart, A. (1961). *The Advanced Theory of Statistics.* Volume II. London: Griffin.

Kiers, H. (1989). *Three-way methods for the analysis of qualitative and quantitative two-way data.* Leiden University: DSWO Press.

Kim, H. (1992). Measures of influence in correspondence analysis. *Journal of Statistical Computation and Simulation, 40*, 201-217.

Kobayashi, R. (1981). *Suryoka Riron Nyumon (Introduction to Quantification Theory).* Tokyo: Nikka Giren.

Komazawa, T. (1978). *Tagenteki Data Bunseki no Kiso (Foundations of Multidimensional Data Analysis).* Tokyo: Asakura Shoten. (in Japanese).

Komazawa, T. (1982). *Suryoka Riron to Data Shori (Quantification Theory and Data Analysis).* Tokyo: Asakura Shoten. (in Japanese).

Komazawa, T. Hashiguchi, K. and Ishizaki, R. (1998). *Pasokon Suryoka Bunseki (Quantification Analysis with Personal Computers).* Tokyo: Asakura Shoten.

Koster, J.T.A. (1989). *Mathematical aspects of multiple correspondence analysis for ordinal variables.* Leiden University: DSWO Press.

Kroonenberg, P.M. (2001). Three-mode correspondence analysis; an illustrated expose. *Actes des XXXIIIemes Journée de Statistiqu,* 101-108.

Kroonenberg, P. M. (2002). Analyzing dependence in large contingency tables: nonsymmetric correspondence analysis and regression with optimal scaling. In Nishisato, S., Baba, Y., Bozdogan, H. and Kanefuji, K. (eds.), *Measurement and Multivariate Analysis.* Tokyo: Springer, 87-96.

Kroonenberg, P.M. and Lombardo, R. (1998). Nonsymmetric correspondence analysis: a tutorial. *Kwantitatieve Methoden, 19,* 57-83.

Kroonenberg, P.M. and Lombardo, R. (1999). Nonsymmetric correspondence analysis: a tool for analyzing contingency tables with a dependence structure. *Multivariate Behavioral Research, 34,* 367-397.

Kzranowski, W.J. (1994). Ordination in the presence of group structure for general multivariate data. *Journal of Classification, 11,* 195-207.

Lancaster, H.O. (1953). A reconciliation of χ^2, considered from metrical and enumerative aspects. *Sankhya, 13,* 1-10.

Lancaster, H.O. (1958). The structure of bivariate distribution. *Annals of Mathematical Statistics, 29,* 719-736.

Lauro, N.C. and D'Ambra, L. (1984). L'analyse non symétrique des correspondances. In Diday, E. al. (eds.), *Data Analysis and Informatics.* Amsterdam: North Holland.

Lauro, N.C. and Decarli, A. (1982). Correspondence analysis and loglinear models in multiway contingency tables study. *Metron,* 1-2, 213-234.

Lauro, N.C. and Siciliano, R. (1988). Correspondence analysis and modeling for contingency tables: symmetric and nonsymmetric approaches. The Third International Workshop on Statistical Modeling. Vienna.

Lawrence, D.R. (1985). Dual scaling of multidimensional data structures. An extended comparison of three methods. Doctoral thesis, University of Toronto.

Lebart, L. (1976). The significance of eigenvalues issued from correspondence analysis. *COMSTAT,* 38-45.

Lebart, L. (1982). L'analyse statistique de réponses libres das les enquêtes socio-economiques. *Consommation Revue de Socio-Economie,* 1, 39-62.

Lebart, L. and Fénelon, J.P. (1971). *Statistique et Informatique Appliqées.* Paris: Dunod.

Lebart, L. and Morineau, A. (1981). Statistical significant criteria in multiple-choice data reduction and visualization. Paper presented at the Annual Meeting of the Psychometric Society, Chapel Hill.

Lebart, L., Morineau, A. and Fénelon, J.P. (1979). *Traitement des Données Statistiques.* Paris: Dunod.

Lebart, L., Morineau, A. and Tabard, N. (1977) *Techniques de la Description Statistique: Méthodes et Logiciels pour l'Analyse des Grands Tableaux.* Paris: Dunod.

Lebart, L., Morineau, A. and Warwick, K.M. (1984). *Multivariate descriptive statistical analysis.* New York: Wiley.

Leclerc, A., Chevalier, A., Luce, D. and Blanc, M. (1985). Analyse des correspondances et modéle logistique: possibilitées et intêrte d'approches complêmentaires. Revue de Statistique Appliqeée, XXXIII, 1.

Legendre, P. and Legendre, L. (1994). *Numerical Ecology*. Amsterdam: North-Holland.

Lehmann, D. and Hurlbert, J. (1972). Are three-point scales always good enough? *Journal of Marketing Research, 9*, 444-446.

Lenoble, F. (1927). A propos des associationsvégétales. *Bull. Soc. bot. Fr., 73*, 873-893.

Le Roux, B. and Rouanet, H. (2004). *Geometric Data Analysis: From Correspondence Analysis to Structured Data*. Dordrecht: Kluwer.

Likert, R. (1932). A technique for the measurement of attitudes. *Archives of Psychology*, No. 140, 44-53.

Lingoes, J.C. (1964). Simultaneous linear regression: An IBM 7090 program for analyzing metric/nonmetric or linear/nonlinear data. *Behavioral Science, 9*, 87-88.

Lingoes, J.C. (1968). The multivariate analysis of qualitative data. *Multivariate Behavioral Research, 3*, 61-94.

Lingoes, J.C. (1973). *The Guttman-Lingoes Nonmetric Program Series*. Ann Arbor: Mathesis Press.

Little, R.A.J. and Rubin, D.B. (1990). The analysis of social science data with missing values. In Fox, J. and Scott Long, T. (eds.), *Modern Methods of Data Analysis*. London: Sage, 374-409.

Lord, F.M. (1958). Some relations between Guttman's principal components of scale analysis and other psychometric theory. *Psychometrika, 23*, 291-296.

Loucks, O.L. (1962). Ordinating forest communities by means of environmental scalars and phytosociological indices. *Ecol. Monogr., 32*, 137-166.

Maeda, T. (1996). Analysis of structured two-way tables and partial correspondence analysis. *Institute of Statistical Mathematics Research Report 86 "Project on Structural Analysis of Multivariate Qualitative Data."* 52-59 (in Japanese).

Maeda, T. (1997). Several features of partial correspondence analysis and its applications. *Institute of Statistical Mathematics Research Report 100. "Project on Structural Analysis of Multivariate Qualitative Data."* 81-89.

Markus, M.T. (1994). *Bootstrap Confidence Regions in Nonlinear Multivariate Analysis*. Leiden: DSWO Press.

Maung, K. (1941a). Measurement of association in contingency tables with special reference to the pigmentation of hair and eye colours of Scottish children. *Annals of Eugenics, 11*,189-223.

Maung, K. (1941b). Discriminant analysis of Tocher's eye colour data for Scottish school children. *Annals of Eugenics, 11*, 64-76.

Mayenga, C. (1997). Dual scaling of sorting data: Effects of limiting categorization on quantification results. Doctoral Thesis, University of Toronto.

McDonald, R.P. (1968). A unified treatment of the weighting problem. *Psychometrika, 33*, 351-381.

McDonald, R.P. (1983). Alternative weights and invariant parameters in optimal scaling. *Psychometrika, 48*, 377-391.

McDonald, R.P., Torii, Y. and Nishisato, S. (1979). Some results on proper eigenvalues and eigenvectors with applications to scaling. *Psychometrika, 44*, 211-227.

McKeon, J.J. (1966). Canonical analysis: some relations between canonical correlation, factor analysis, discriminant function analysis and scaling theory. *Psychometric Monograph No. 13.*

Meulman, J. (1982). *Homogeneity analysis of incomplete data.* Leiden University: DSWO Press.

Meulman, J. (1984). Correspondence analysis and stability. Research Report 84-01, Department of Data Theory, Leiden University.

Meulman, J. (1986). *A distance approach to nonlinear multivariate analysis.* Leiden University: DSWO Press.

Meulman, J. (1998). Review of Kzranowski, W.J. "Multivariate Analysis: Part I. Distributions, Ordinations, and Inference." *Journal of Classification, 15*, 297-298.

Meulman, J. (2003). Prediction and classification in nonlinear data analysis: Something old, something new, something borrowed and something blue. *Psychometrika, 68*, 493-517.

Michailidis, G. and de Leeuw, J. (1998). The Gifi system of descriptive multivariate analysis. *Statistical Science, 13*, 307-336.

Millones, O. (1991). Dual scaling in the framework of the association and correlation models under maximum likelihood. Doctoral Thesis, University of Toronto.

Mizuta, M. (2002). Relative projection pursuit. In Sokotowski, A. and Jajuga, K. (eds.), *Data Analysis, Classification, and Related Methods.* Crakow University of Economics.

Mizuta, M. and Hiro, S. (2004). Relative projection pursuit and its applications. In Banksw, D., House, L., McMorris, F.R., Arabie, P. and Gaul, W. (eds.P), *Classification, Clustering, and Data Mining Application.* Springer, 1230139.

Morimoto, E. (1997a). Suryoka-riron no keisei (The formation of Hayashi's quantification theory). *Journal of History of Science, Japan. 36*, 85-95.

Morimoto, E. (1997b). The formation of Hayashi's quantification theory. In Knobloch, E., Mawhin, J. and Demirov, S. (eds.), *Studies in History of Mathematics Dedicated to A.P.Youschkevitch.* Liége: Prepols, 319-324.

Morimoto, E. (1999). Suryoka-riron no fukyu - Riron keisei-go no 1950 nen-dai kara 1970 nen made no tenkai (The dissemination of Hayashi's quantification methods: The development from the 1950s when these methods were formulated to 1970). *Journal of History of Science, Japan. 38*, 129-141.

Mosier, C.I. (1946). Machine methods in scaling by reciprocal averages. *Proceedings, Research Forum.* Endicath, N.Y.: International Business Corporation. 35-39.

Mosteller, F. (1949). A theory of scalogram analysis using noncumulative types of items. Report No.9, Laboratory of Social Relations, Harvard University.

Mucha, H.J. (2002). An intelligent clustering technique based on dual scaling. In Nishisato, S., Baba, Y., Bozdogan, H. and Kanefuji, K. (eds.), *Measurement and Multivariate Analysis*. Tokyo: Springer, 37-46.

Murtagh, F. (2005). *Correspondence Analysis and Data Coding with R and Java*. Boca Raton: Chapman and Hall.

Myers, I.B. (1962). *The Myers-Briggs Type Indicator Manual*. Princeton: Educational Testing Service.

Nagy, P. (1984 ed.). *The Representation of Cognitive Structures*. Toronto: Department of Measurement, Evaluation and Computer Applications, the Ontario Institute for Studies in Education.

Nakache, J.P. (1982). *Exercises Commentés de Mathématiques pour l'Analyse des Données*. Paris: Dunod.

Nakayama, T., Naito, K. and Fujikoshi, Y. (1998). Stability of correspondence analysis and its alternative using Hellinger distance for contingency table. *International Journal of Mathematics and Statistical Science, 7*, 97-119.

Nishisato, S. (1971). Analysis of variance through optimal scaling. *Proceedings of the First Canadian Conference on Applied Statistics*. Montreal: Sir George Williams University Press, 306-316.

Nishisato, S. (1972). Analysis of variance of categorical data through selective scaling. *Proceedings of the 20th International Congress of Psychology*. Tokyo, 279.

Nishisato, S. (1975). *Oyo Shinri Shakudoho: Shitsuteki Data no Bunseki to Kaishaku (Applied Psychological Scaling: Analysis and Interpretation of Qualitative Data)*. Tokyo: Seishin Shobo. (in Japanese)

Nishisato, S. (1978). Optimal scaling of paired comparison and rank order data: an alternative to Guttman's formulation. *Psychometrika, 43*, 263-271.

Nishisato, S. (1980a). *Analysis of Categorical Data: Dual Scaling and Its Applications*. Toronto: University of Toronto Press.

Nishisato, S. (1980b). Dual scaling of successive categories data. *Japanese Psychological Research, 22*, 134-143.

Nishisato, S. (1982). *Shitsuteki Data no Suryoka: Sotsui Shakudo ho to Sono Oyo (Quantifying Qualitative Data: Dual Scaling and Its Applications)*. Tokyo: Asakura Shoten. (in Japanese).

Nishisato, S. (1984a). Forced classification: A simple application of a quantification technique. *Psychometrika, 49*, 25-36.

Nishisato, S. (1984b). Dual scaling by reciprocal medians. it Estratto Dagli Atti della XXXII riunione Scientifica. Sorrento, 141-147.

Nishisato, S. (1986a). Generalized forced classification for quantifying categorical data. In Diday, E. et al. (eds.), *Data Analysis and Informatics*. Amsterdam: North-Holland, 351-362.

Nishisato, S. (1986b). Multidimensional analysis of successive categories. In J. de Leeuw, W. Heiser, J. Meulman, and F. Critchley (eds.), *Multidimensional Data Analysis*. Leiden: DSWO Press, 249-250.

Nishisato, S. (1986c). *Quantification of Categorical Data: A Bibliography 1975-1986*. Toronto: MicroStats.

Nishisato, S. (1987). Robust techniques for quantifying categorical data. In MacNeil, I.B. and Umphrey, G.J.(eds.), *Foundations of Statistical Inference*. Dordrecht: D. Reidel Publishing Company, 209-217.

Nishisato, S. (1988a). Forced classification procedure of dual scaling: its mathematical properties. In Bock, H.H. (ed.), *Classification and Related Methods*. Amsterdam: North-Holland, 523-532.

Nishisato, S. (1988b). Market segmentation by dual scaling through generalized forced classification. In Gaul, W. and Schader, M. (eds.), *Data, Expert Knowledge and Decisions*. Berlin: Springer-Verlag, 268-278.

Nishisato, S. (1988c). Effects of coding on dual scaling. A paper presented at the Annual Meeting of the Psychometric Society, University of California, Los Angeles.

Nishisato, S. (1991). Standardizing multidimensional space for dual scaling. In the *Proceedings of the 20th Annual Meeting of the German Operations Research Society*. Hohenheim University, 584-591.

Nishisato, S. (1993). On quantifying different types of categorical data. *Psychometrika, 58,* 617-629.

Nishisato, S. (1994). *Elements of Dual Scaling: An Introduction to Practical Data Analysis*. Hilsdale, N.J.: Lawrence Erlbaum Associates.

Nishisato, S. (1996). Gleaning in the field of dual scaling. *Psychometrika, 61,* 559-599.

Nishisato, S. (1999). Data types and information: beyond the current practice of data analysis. In Decker, R. and Gaul, W. (eds.), *Classification and Information Processing at the Turn of the Millennium*. Heidelberg: Springer-Verlag, 40-51.

Nishisato, S. (2000a). Data analysis and information: Beyond the current practice of data analysis. Decker, R. and Gaul, W. (eds.), *Classification and Information Processing at the Turn of the Millennium*, Heidelberg: Springer-Verlag, 40-51.

Nishisato, S. (2000b). A characterization of ordinal data. In Gaul, W., Opitz, O. and Schader, M. (eds.), *Data Analysis; Scientific Modeling and Practical Applications*. Heidelberg: Springer-Verlag, 285-298.

Nishisato, S. (2002). Differences in data structures between continuous and categorical variables from dual scaling perspectives, and a suggestion for a unified mode of analysis. *Japanese Journal of Sensory Evaluation, 6,* 89-94 (in Japanese).

Nishisato, S. (2003a). Geometric perspectives of dual scaling for assessment of information in data. In Yanai, H., Okada, A., Shigemasu, K., Kano, Y. and Meulman, J. J. (eds.), *New developments in psychometrics*. Springer-Verlag, 453-462.

Nishisato, S. (2003b). A proposal to mathematical statistics from a psychometrician: A framework for multidimensional data analysis. In Kano, Y. and

Chino, N. (eds.), *Symposium on Bridging Between Mathematical Statistics and Psychometrics*. Osaka University, 103-110 .

Nishisato, S. (2003c). Total information in multivariate data from dual scaling perspectives. *The Alberta Journal of Educational Research, XLIX*, 244-251.

Nishisato, S. (2004). A unified framework for multidimensional data analysis from dual scaling perspectives: Another look and some suggestions. *Japanese Journal of Sensory Evaluation, 8*, 4-7. (in Japanese).

Nishisato, S. (2005a). New framework for multidimensional data analysis. In Weihs, C. and Gaul, W. (eds.), *Classification - the Ubiquitous Challenge*. Heidelberg: Springer, 280-287.

Nishisato, S. (2005b). On the scaling of ordinal measurement: a dual scaling perspective. In Maydeu-Olivares, A. and McArdle, J.J.(eds.), *Contemporary Psychometrics*. Mahwah: Lawrence Erlbaum, 479-507.

Nishisato, S. (2005c). Interpretation of data in multidimensional space. Paper presented at the 70th Annual Meeting of the Psychometric Society, Tilburg.

Nishisato, S. (2006). *Multidimensional Nonlinear Descriptive Analysis*. London: Chapman and Hall/CRC.

Nishisato, S. and Ahn, H. (1995). When not to analyze data: Decision making on missing responses in dual scaling. *Annals of Operations Research, 55*, 361-378.

Nishisato, S. and Arri, P.S. (1975). Nonlinear programming approach to optimal scaling of partially ordered categories. *Psychometrika, 40*, 525-548.

Nishisato, S. and Baba, Y. (1999). On contingency, projection and forced classification of dual scaling. *Behaviormetrika, 26*, 207-219.

Nishisato, S. and Clavel, J.G. (2002). A note on between-set distances in dual scaling and correspondence analysis. *Behaviormetrika, 30,* 87-98.

Nishisato, S. and Gaul, W. (1990). An approach to marketing data analysis. *Journal of Marketing Research, 27*, 354-360.

Nishisato, S. and Hemsworth, D. (2002). Quantification of ordinal variables: A critical inquiry into polychoric and canonical correlation. In Baba, Y., Hayter, A.J., Kanefuji, K. and Kuriki, S. (eds.), *Recent advances in statistical research and data analysis*. Tokyo: Springer, 49-84.

Nishisato, S. and Inukai, Y. (1972). Partially optimal scaling of items with ordered categories. *Japanese Psychological Research, 14*, 109-119.

Nishisato, S. and Lawrence, D.R. (1989). Dual scaling of multiway data matrices: several variants. In Coppi, R. and Bolasco, S. (eds.), *Multiway Data Analysis*. Amsterdam: Elsevier Science Publishers, 317-326.

Nishisato, S. and Nishisato, I. (1984). *An Introduction to Dual Scaling*. Toronto: MicroStats.

Nishisato, S. and Nishisato, I. (1994). *Dual Scaling in a Nutshell*. Toronto: MicroStats.

Nishisato, S. and Sheu, W.J. (1980). Piecewise method of reciprocal averages for dual scaling of multiple-choice data. *Psychometrika, 45*, 467-478.

Nishisato, S. and Sheu, W.J. (1984). A note on dual scaling of successive categories data. *Psychometrika, 49*, 493-500.

Noma, E. (1982). The simultaneous scaling of cited and citing articles in a common space. *Scientometrics, 4*, 205-231.

Odondi, M.J. (1997). Multidimensional analysis of successive categories (rating) data by dual scaling. Doctoral thesis, University of Toronto.

Ohsumi, N., Lebart, L., Morineau, A., Warwick, K.M. and Baba, Y. (1994). *Descriptive Multivariate Analysis.* Tokyo: Nikkagiren. (in Japanese).

Okada, A. and Imaizumi, T. (1997). Asymmetric multidimensional scaling of two-mode three-way proximities. *Journal of Classification, 14*, 195-224.

Okada, A. and Imaizumi, T. (2000). Two-mode three-way asymmetric multidimensional scaling with constraints on asymmetry. In Decker, R. and Gaul, W. (eds.), *Classification and Information Processing at the Turn of the Millennium.* Springer: Berlin, 52-59.

Okada, A. and Imaizumi, T. (2002). Multidimensional scaling with different orientations of symmetric and asymmetric relationships. In Nishisato, S., Baba, Y., Bozdogan, H. and Kanefuji, K. (eds.), *Measurement and Multivariate Analysis.* Springer: Tokyo, 97-106.

Okada, A. and Imaizumi, T. (2003). Two-mode three-way nonmetric multidimensional scaling with different directions of asymmetry for different sources. In Yanai, H., Okada, A., Shigemasu, K., Kano, Y. and Meulman, J. (eds.). *New Developments in Psychometrics.* Springer: Tokyo, 495-502.

Okada, A. and Imaizumi, T. (2005). External analysis of two-mode three-way asymmetric multidimensional scaling. In Weihs, C. and Gaul, W. (eds.), *Classification - the Ubiquitous Challenge.* Springer: Berlin, 288-295.

Okada, A., Imaizumi, T. and Inoue, H. (2005). Asymmetric multidimensional scaling of relationships among managers of a firm. In Baier, D., Decker, R. and Schmidt-Thieme, L. (eds.), *Data Analysis and Decision Support.* Springer: Berlin, 100-107.

Okamoto, Y. (1995). Unfolding by the criterion of the fourth quantification method. *Journal of Behaviormetrics, 22*, 126-134 (In Japanese with English abstract).

Oksanen, J. and Ahti, T. (1982). Lichen-rich pine forest vegetation in Finland. *Ann. Bot. Fennici, 19*, 275-301.

Orlóci, L. (1978). *Multivariate Analysis in Vegitations Research.* 2nd Edition. The Hague: Junk.

Pearson, K. (1901). On lines and planes of closest fit to systems of points in space. *Philosophical Magazines and Journal of Science, Series 6, 2,*559-572.

Pearson, K. (1904). Mathematical contribution to the theory of evolution. XIII. On the theory of contingency and its relation to association and normal correlation. *Drapers' Company Research Memoires, Biometric Series, 1,*1-35.

Perreault, W. and Young, F.W. (1980). Alternating least squares optimal scaling: analysis of nonmetric data in marketing research. *Journal of Marketing Research, 17*, 1-13.

Phillips, J.P.N. (1971). A note on the presentation of ordered metric scaling.

British Journal of Mathematical and Statistical Psychology, 4, 239-250.

Pierce, J.R. (1961). *Symbols, Signals and Noise: The Nature and Process of Communication.* New York: Harper and Rowe Publishers.

Poon, W.P. (1977). Transformations of data matrices in optimal scaling. Master Thesis, University of Toronto.

Prentice, I.C. (1977). Non-metric ordination methods in ecology. *Journal of Ecology, 65,* 85-94.

Ramensky, L.G. (1930). Zür Methodik der vegleichenden und Ordnung von Pfanzenlisten und anderen Objecten, die durch mehrere, verschiedenartig wirkende Factoren bestimmt werden. *Beitr. Biol. Pft., 18,* 269-304.

Rao, C.R. (1952). *Advanced Statistical Methods for Biometric Research.* New York: Wiley.

Rao, C.R. (1995a). A review of canonical coordinates and an alternative in correspondence analysis using Hellinger distance. *Qüestiió, 19,* 23-63.

Rao, C.R. (1995b). The use of Hellinger distance in graphical display of contingency table data. *NTProSta3, 3,* 143-161.

Richardson, M. and Kuder, G. F. (1933). Making a rating scale that measures. *Personnel Journal, 12,* 36-40.

Ritov, Y. and Gilula, Z. (1993). Analysis of contingency tables by correspondence models subject to order constraints. *Journal of the American Statistical Association, 88,* 1380-1387.

Ross, J. and Cliff, N. (1960). A generalization of the interpoint distance model. *Psychometrika, 29,* 167-176.

Rouanet, H. and Le Roux, B. (1993). *Analyse des Données Multidimensionelles.* Paris: Dunod.

Rowe, J.S. (1956). Uses of undergrowth plant species in forestry. *Ecology, 37,* 461-473.

Sachs, J. (1994). Robust dual scaling weights with Tukey's biweight. *Applied Psychological Measurement, 18,* 301-309.

Saito, T. (1980). *Tajigen Shakudo Ho (Multidimensional Scaling).* Tokyo: Asakura Shoten. (in Japanese).

Saporta, G. (1975). Liaisons entre plusieurs ensembles de variables et codage de donées qualitatives. Doctoral Thesis, L'Université Piérre et Marie Curie, Paris VI, France.

Saporta, G. (1976). Discriminant analysis when all the variables are nominal: A stepwise method. Paper presented at the Symposium on Optimal Scaling, Annual Meeting of the Psychometric Society, Murray Hill.

Saporta, G. (1979). *Theories et Méthodes de la Statistique.* Paris:Dunod.

Saporta, G. (1990). *Probabilit'és, Analyse des Données et Statistique.* Paris: Technip.

Schmidt, E. (1907). Zür Theorie der linearen und nichtlinearen Integral-gleichungen. Esster Teil. Entwickelung willkürlicher Functionen nach Systemaen vorgeschriebener (On theory of linear and nonlinear integral equations. Part one. Development of arbitrary functions according to prescribed systems). *Mathematische Annalen, 63,* 433-476.

Schönemann, P.H. (1970). On metric multidimensional unfolding. *Psychometrika, 35,* 167-176.

Schönemann, P.H., Bock, R.D. and Tucker, L.R. (1965). Some notes on a theorem by Eckart and Young. *Research Memorandum No.25,* The Psychometric Laboratory, University of North Carolina.

Schönemann, P.H. and Wang, M.M. (1972). An individual difference model for the multidimensional analysis of preference data. *Psychometrika, 38,* 275-309.

Schriever, B.F. (1983). Scaling of order dependent categorical variables with correspondence analysis. *International Statistical Review, 51,* 225-238.

Shaw, D.G., Huffman, M.D. and Haviland, M.G. (1987). Grouping continuous data in discrete intervals: Information loss and recovery. *Journal of Educational Measurement, 24,* 167-173.

Sheskin, D.J. (1997). *Handbook of parametric and nonparametric procedures.* Boca Raton, Florida: CRC Press.

Siciliano, R., Mooijaart, A. and van der Heijden, P.G.M. (1990). Non-symmetric correspondence analysis by maximum likelihood. Technical Report PRM 05-90. Leiden University.

Sixtl, F. (1973). Probabilistic unfolding. *Psychometrika, 38,* 235-248.

Skinner, H.A. and Sheu, W.J. (1982). Dimensional analysis of rank-order and categorical data. *Applied Psychological Measurement, 6,* 41-45.

Slater, P. (1960). Analysis of personal preferences. British Journal of Statistical Psychology, 3, 119-135.

Stefansky, W. and Kaiser, H.F. (1973). Note on discrete approximations. *Journal of the American Statistical Association, 68,* 232-234.

Stevens, S.S. (1951). Mathematics, measurement, and psychophysics. In Stevens, S.S. (ed.), *Handbook of Experimental Psychology.* New York: Wiley.

Takane, Y. (1980a). *Tajigen Shakudo ho (Multidimensional Scaling).* Tokyo: University of Tokyo Press. (in Japanese).

Takane, Y. (1980b). Analysis of categorizing behavior. *Behaviormetrika, 8,* 75-86.

Takane, Y. (1981). Multidimensional successive categories scaling: a maximum likelihood method. *Psychometrika, 46,* 9-28.

Takane, Y. (1984). Multidimensional scaling of sorting data. In Chaubey, Y.P. and Dwivedi, T.D. (eds.) *Topics in Applied Statistics.* Montreal: Concordia University Press, 659-666.

Takane, Y. and Hwang, H. (2005). Regularization methods in multivariate analysis. Paper presented at the 70th Annual Meeting of the Psychometric Society, Tilburg.

Takane, Y. and Shibayama, T. (1991). Principal component analysis with external information on both subjects and variables. *Psychometrika, 56,* 97-120.

Takane, Y., Yanai, H. and Mayekawa, S. (1991). Relationships among several methods of linearly constrained correspondence analysis. *Psychometrika, 56,*

667-684.

Takane, Y., Young, F.W. and de Leeuw, J. (1980). An individual differences additive model: an alternating least squares method with optimal scaling features. *Psychometrika, 45*. 183-209.

Takeuchi, K. and Yanai, H. (1972). *Tahenryou Kaiseki no Kiso (Foundations of Multivariate Analysis)*. Tokyo: Toyo Keizai Shimpo sha. (in Japanese).

Tanaka, Y. (1978). Some generalized method of optimal scaling and their asymptotic theories: the case of multiple responses-multiple factors. *Annals of Institute of Statistical Mathematics, 30*, 329-348.

Tanaka, Y. (1983). Sensitivity analysis in the methods of quantification. *Mathematical Sciences No.245*, 32-37 (in Japanese).

Tanaka, Y. (1984a). Sensitivity analysis in Hayashi's third method of quantification. *Behaviormetrika, 16*, 31-44.

Tanaka, Y. (1984b). Sensitivity analysis for quantification theory and its applications. *Quality, 24*, 337-345. (in Japanese).

Tanaka, Y. (1992). Sensitivity analysis in multivariate methods. *Japanese Journal of Behaviormetrics, 19*, 3-17.

Tanaka, Y. and Asano, C. (1978). On some basic properties of optimal scaling for ordered categories. *Research Report 84*. Research Institute of Fundamental Information Science, Kyushu University.

Tanaka, Y., Asano, C. and Kodake, K. (1978). Application of nonlinear programming techniques to optimal scaling for ordered categories. *Research Report 85*. Research Institute of Fundamental Information Science, Kyushu University.

Tanaka, Y., Asano, C. and Kubota, N. (1978). A generalized method of optimal scaling for multiple responses with ordered categories. *Research Report 86*. Institute of Fundamental Information Science, Kyushu University.

Tanaka, Y. and Kodake, K. (1980). Computational aspects of optimal scaling for ordered categories. *Behaviormetrika, 7*, 35-46.

Tanaka, Y. and Morikawa, T. (1980). Optimal scaling for multi-item ordered categories: A branch and bound procedure. *Reports of Research and Development, Volume. 1*. Okayama University Computer Center, 65-83.

Tanaka, Y, and Tarumi, T. (1985). Computational aspect of sensitivity analysis in multivariate methods. *Technical Report No.12*, Okayama Statisticians Group.

Tanaka, Y. and Tarumi, T. (1988a). Sensitivity analysis in Hayashi's second method of quantification. *Journal of Japanese Statistical Society, 16*, 37-57.

Tanaka, Y. and Tarumi, T. (1988b). Outliers and influential observations in quantification theory. In Diday, E. et al. (eds.), *Recent Developments in Clustering and Data Analysis*. London: Academic Press, 281-293.

Tanaka, Y. and Tarumi, T. (1988c). Sensitivity of the geometrical representation obtained by correspondence analysis to small changes of data. In Das Gupta, S. and Ghosh, J.K. (eds.), *Advances in Multivariate Statistical Analysis*. Indian Statistical Institute, 499-511.

Tarumi, T. (1986). Sensitivity analysis of descriptive multivariate methods formulated by the generalized singular value decomposition. *Math. Japon.,* *31*, 957-977.

Tarumi, T. and Tanaka, Y. (1986). Statistical software SAM - Sensitivity analysis in multivariate methods. *COMSTAT*, 351-356.

Tateneni, K. and Browne, M.W. (2000). A noniterative method of joint correspondence analysis. it Psychometrika, 65, 157-165.

Tchuproff, A.A. (1925). *Grundbegriffe und Grund Problem der Korrelationstheorie.* Leipzig: Teubner.

Teil, H. (1975). Correspondence factor analysis: An outline of its method. *Mathematical Geology, 7*, 3-12.

Tenenhaus, M. (1982). Multiple correspondence analysis and duality schema: a synthesis of different approaches. *Metron, XL*, 289-302.

Tenenhaus, M. (1994). *Méthodes Statistiques en Gestion.* Paris: Dunod.

Tenenhaus, M. and Young, F.W. (1985). An analysis and synthesis of multiple correspondence analysis, optimal scaling, dual scaling, homogeneity analysis and other methods for quantifying categorical data. *Psychometrika, 50*, 91-119.

Ter Braak, C.J.F. (1983). Principal component biplots and alpha and beta diversity. *Ecology, 64*, 454-462.

Ter Braak, C.J.F. (1985). CANOCO: A FORTRAN program for canonical correspondence analysis and detrended correspondence analysis. Wageningen: IWIS-TNO.

Ter Braak, C.J.F. (1986). Canonical correspondence analysis: a new eigenvector technique for multivariate direct gradient analysis. *Ecology, 67*, 1167-1179.

Ter Braak, C.J.F. (1987). Ordination. In Jongman, R.H.G., Ter Braak, C.J.F. and Van Tongeren, C.F.R. (eds.), *Data Analysis in Community and Landscape Ecology.* Wageningen: Pudoc, 91-173.

Ter Braak, C.J.F.(1988). Partial canonical correspondence analysis. In Bock, H.H. (ed.), *Classification and Related Methods of Data Analysis.* Amsterdam: North-Holland, 551-558.

Torgerson, W.S. (1952). Multidimensional scaling. I. Theory and method. *Psychometrika, 17*, 401-419.

Torgerson, W.S. (1958). *Theory and Methods of Scaling.* New York: Wiley.

Torres-Lacomba, A. and Greenacre, M.J. (2002). Dual scaling and correspondence analysis of preferences, paired comparisons and ratings. *International Journal of Research in Marketing, 19*, 401-405.

Tsujitani, M. (1984). A review of the analysis of ordered categorical data. I. *Japanese Journal of Behaviormetrics, 11*, 22-38 (in Japanese).

Tsujitani, M. (1985). A review of the analysis of ordered categorical data. II. *Japanese Journal of Behaviormetrics, 13*, 33-43 (in Japanese).

Tsujitani, M. (1986). A review of the analysis of ordered categorical data. III. *Japanese Journal of Behaviormetrics, 14*, 47-59 (in Japanese).

Tsujitani, M. (1987). Maximum likelihood methods for association models in ordered categorical data. *Behaviormetrika, 22*, 61-67.

Tsujitani, M. (1988a). Optimal scaling for association models when category scores have a natural ordering. *Statistics and Probability Letters, 6*, 175-180.

Tsujitani, M. (1988b). Maximum likelihood methods for association models in ordered categorical data: multi-way case. *Behaviormetrika, 23*, 85-91.

Tsujitani, M. and Koch, G.G. (1988). Loglinear models approach to the analysis of association in multiway cross-classifications having ordered categories. *Reports of Statistical Application Research, JUSE, 35*, 1-10.

Tucker, L. R. (1960). Intra-individual and inter-individual multidimensionality. In H. Gulliksen and S. Messick (eds.), *Psychological Scaling*. New York: Wiley.

Underhill, L.G. (1990). The coefficient of variation biplot. *Journal of Classification, 7*, 41-56.

van Buuren, S. (1990). *Optimal scaling of time series*. Leiden University: DSWO Press.

van Buuren, S. and van Rijckevorsel, J.L.A. (1992). Imputation of missing categorical data by maximizing internal consistency. *Psychometrika, 57*, 567-580.

van de Geer, J.P. (1993). *Multivariate Analysis of Categorical Data: Applications*. Newbury Park: Sage Publications.

van de Velden, M. (2000). Dual scaling and correspondence analysis of rank order data. In Heijmans, R.D.H., Pollock, D.S.G., and Satorra, A. (eds.), *Innovations in multivariate statistical analysis*. Dordrecht: Kluwer Academic Publishers.

van der Burg, E. (1988). *Nonlinear canonical correlation and some related techniques*. Leiden University: DSWO Press.

van der Heijden, P.G.M. (1987). *Correspondence analysis of longitudinal categorical data*. Leiden University: DSWO Press.

van der Heijden, P.G.M., de Falgruerolles, A. and de Leeuw, J. (1989). A combined approach to contingency table analysis with correspondence analysis and log-linear analysis (with discussion). *Applied Statistics, 38*, 249-292.

van der Heijden, P.G.M. and de Leeuw, J. (1985). Correspondence analysis used complimentary to loglinear analysis. *Psychometrika, 50*, 429-447.

van der Heijden, P.G.M. and Escofier, B. (1989). Multiple correspondence analysis with missing data. Unpublished manuscript, Department of Psychometrics and Research Methods, Leiden University.

van Os, B.J. (2000). *Dynamic Programming for Partitioning in Multivariate Data Analysis*. Leiden: Universal Press.

van Rijckevorsel, J. (1987). *The application of fuzzy coding and horseshoes in multiple correspondence analysis*. University of Leiden: DSWO Press.

van Rijckevorsel, J.and de Leeuw, J. (1988). *Component and Correspondence Analysis: Dimension Reduction by Functional Approximation*. New York: Wiley.

Verboon, P. (1994). *A robust approach to nonlinear multivariate analysis*. Leiden

University: DSWO Press.

Weingarden, P. and Nishisato, S. (1986). Can a method of rank ordering reproduce paired comparison? An analysis by dual scaling (correspondence analysis). *Canadian Journal of Marketing Research, 5,* 11-18.

Weller, J.C. and Romney, A.K. (1990). *Metric Scaling: Correspondence Analysis.* Newbury: Sage Publications.

Whittaker, R.H. (1948). A vegetation analysis of the Great Smoky Mountains. Ph.D. thesis, University of Illinois.

Whittaker, R. H. (1966). Forest dimensions and production in the Great Smoky Mountains. *Ecology, 47,* 103-121.

Whittaker, R.H. (1967). Gradient analysis of vegetation. *Biological Review, 42,* 206-264.

Whittaker, R.H. (1978a). Direct gradient analysis. In Whittaker, R.H. (ed.), *Ordination of Plant Community.* The Hague: Junk, 7-50.

Whittaker, R.H. (1978b) (ed.), *Ordination of Plant Community.* The Hague: Junk.

Whittaker, J. (1989). Discussion of :A combined approach to contingency table analysis using correspondence analysis and log-linear analysis" by P.G.M. van der Heijden, A. de Falguerolles and J. de Leeuw. *Applied Statistics, 38,* 278-279.

Wiley, D.E. (1967). Latent partition analysis. *Psychometrika, 32,* 183-194.

Wilks, S.S. (1938). Weighting system for linear function of correlated variables when there is no dependent variable. *Psychometrika, 3,* 23-40.

Williams, E. J. (1952). Use of scores for the analysis of association in contingency tables. *Biometrika, 39,* 274-289.

Williams, O.D. and Grizzle, J.E. (1972). Analysis of contingency tables having ordered response categories. *Journal of the American Statistical Association, 67,* 55-63.

Yamada, F. (1994). A review of "Nishisato, S. 'Elements of Dual Scaling: An Introduction to Practical Data Analysis.' Hillsdale: Lawrence Erlbaum, 1994." *Japanese Journal of Behaviormetrics, 21,* 42-43.

Yamada, F. (1996). Dual scaling for data with many missing responses. *Japanese Journal of Behaviormetrics, 23,* 95-103.

Yamada, F. and Nishisato, S. (1993). Several mathematical properties of dual scaling as applied to dichotomous item category data. *Japanese Journal of Behaviormetrics, 20,* 56-63. (in Japanese)

Yamakawa, A., Ichihashi, H. and Miyoshi, T. (1998). Multiple correspondence analysis using L_s−norm and its application to an analysis of senior simulation. In *Proceedings of the 2nd Japan-Australia Joint Workshop on Intelligent and Evolutionary Systems,* 99-106.

Yamanishi, K. and Li, H. (2002). Mining open answers in questionnaire data. *IEEE Intelligent Systems, 17,* 58.

Yanai, H. (1986). Some generalizations of correspondence analysis in terms of Projection operators. In Diday, E.L. et al. (eds.), *Data Analysis and*

Informatics IV. Amsterdam: North-Holland, 193-207.

Yanai, H. (1987). Partial correspondence analysis and its properties. In Hayashi, C. et al. (eds.), *Recent Developments in Clustering and Data Analysis.* Tokyo: Academic Press, 259-266.

Yanai, H. (1988). Partial correspondence analysis and its properties. In Hayashi, C., Jambu, M., Diday, E. and Ohsumi, N. (eds.), *Recent Developments in Clustering and Data Analysis.* Boston: Academic Press, 259-266.

Yanai, H. (1992). *Tahenryo Data Kaisekihou (Multivariate Data Analysis).* Tokyo: Asakura Shoten. (in Japanese)

Yanai, H. and Maeda, T. (2002). Partial multiple correspondence analysis. In Nishisato, S., Baba, Y., Bozdogan, K. and Kanefuji, K. (eds.), *Measurement and Multivariate Analysis.* Tokyo: Springer, 57-68.

Yoshizawa, T. (1975). Models for quantification techniques in multiple contingency tables: the theoretical approach. *Japanese Journal of Behaviormetrics, 3,* 1-11. (in Japanese).

Yoshizawa, T. (1976). A generalized definition of interaction and singular value decomposition of multiway arrays. *Japanese Journal of Behaviormetrics, 4,* 32-43. (in Japanese).

Yoshizawa, T. (1977). Structure of multiway data and population spaces. Doctoral Thesis, Tokyo University.

Young, F.W. (1981). Quantitative analysis of qualitative data. *Psychometrika, 46,* 357-388.

Young, F.W., de Leeuw, J. and Takane, Y. (1976). Regression with qualitative and quantitative variables: an alternating least squares method with optimal scaling features. *Psychometrika, 41,* 505-529.

Young, F.W. , Takane, Y. and de Leeuw, J. (1978). The principal components of mixed measurement level multivariate data: an alternating least squares method with optimal scaling features. *Psychometrika, 43,* 279-281.

Author index

Subject index

T - #0374 - 071024 - C24 - 234/156/15 - PB - 9780367390648 - Gloss Lamination